To John & Anne,
in continuing friendship
& with much affection

from David

June 2007

Science and the Indian Tradition

During the late nineteenth and early twentieth centuries, India experienced an intellectual renaissance that owed as much to the influx of new ideas from the West as to traditional religious and cultural insights. This book examines the effects of the introduction of Western science in India, and the relationship between Indian traditions of thought and Western science. It charts the early development of science in India, its role in the secularization of Indian society, and the subsequent reassertion, adaptation and rejection of traditional modes of thought. It looks at the detailed beliefs of Indian scientists, including Jagadish Chandra Bose, S. N. Bose and P. C. Roy, and reflects upon how individual scientists could accept particular religious beliefs such as reincarnation, cosmology, miracles and prayer. It discusses some of the adaptations of traditional Indian beliefs with insights from Western science, in particular the place of science within the philosophy of Rabindranath Tagore and the 1930 discussions between Einstein and Tagore on the nature of reality. It is argued that the Hindu, Muslim and Christian philosophical and religious traditions have nothing to fear from scientific theories such as evolution and a unified field theory; indeed they may be mutually compatible. Overall, this book provides a detailed assessment of the results of the introduction of Western science into India, and will be of interest to scholars of Indian history and philosophy, historians of science and those interested in the interactions between Western and Indian traditions of intellectual thought.

David L. Gosling is the Principal of Edwardes College in the University of Peshawar, Pakistan, and he also teaches ecology in the University of Cambridge, where he was the first Spalding Fellow at Clare Hall. He has been the Director of Church and Society of the World Council of Churches, and is the author of *Religion and Ecology in India and Southeast Asia*.

India in the Modern World

1 **Privatisation in India**
  Challenging economic orthodoxy
  T. T. Ram Mohan

2 **India – from Regional to World Power**
  Ashok Kapur

3 **Science and the Indian Tradition**
  When Einstein met Tagore
  David L. Gosling

Science and the Indian Tradition

When Einstein met Tagore

David L. Gosling

Routledge
Taylor & Francis Group
LONDON AND NEW YORK

First published 2007
by Routledge
2 Park Square, Milton Park, Abingdon, Oxon OX14 4RN

Simultaneously published in the USA and Canada
by Routledge
270 Madison Ave, New York, NY 10016

Routledge is an imprint of the Taylor & Francis Group, an informa business

© 2007 David L. Gosling

Typeset in Times New Roman by
Newgen Imaging Systems (P) Ltd, Chennai, India
Printed and bound in Great Britain by
MPG Books Ltd, Bodmin

All rights reserved. No part of this book may be reprinted or
reproduced or utilised in any form or by any electronic,
mechanical, or other means, now known or hereafter
invented, including photocopying and recording, or in any
information storage or retrieval system, without permission in
writing from the publishers.

British Library Cataloguing in Publication Data
A catalogue record for this book is available
from the British Library

Library of Congress Cataloging in Publication Data
Gosling, David L., 1939–
    Science and the Indian tradition : when Einstein met Tagore / David L. Gosling.
        p. cm. – (India in the modern world series ; 3)
    Includes bibliographical references and index.
    1. Science – India – History – 19th century. 2. Science – India – History – 20th century. 3. India – Civilization – European influences. 4. India – Intellectual life – 19th century. 5. India – Intellectual life – 20th century. 6. Religion and science – India – History – 19th century. 7. Religion and science – India – History – 20th century. I. Title.
    Q127.I4G77 2007
    303.48′3095409041–dc22   2006034941

ISBN10: 0–415–40209–3 (hbk)
ISBN10: 0–203–96188–9 (ebk)

ISBN13: 978–0–415–40209–5 (hbk)
ISBN13: 978–0–203–96188–9 (ebk)

This book is dedicated to Professor R. Ninian Smart and Libushka Smart

The God of humanity has arrived at the gates of the ruined temple of the tribe.
Rabindranath Tagore, from *Gitāñjali*

Contents

List of illustrations   xi
Acknowledgements   xiii

1 Introduction   1

Science in India 1
Science and the environment 3
Society and tradition 5
Science and religion 7
Einstein meets Tagore 8
Methodology 8
Conclusion 10

2 Science in India's intellectual renaissance   12

Macaulay's Minute 12
The secularization of belief 14
The Reformers 17
The Ramakrishna Mission 18
Raja Yoga 20
Vivekananda and reincarnation 21
Aurobindo Ghose 23
Rejecting the past 24
Science and the Muslims 25
Rabindranath Tagore 27
Conclusion 30

3 Tradition redefined   32

A temple vignette 32
Vedāntic origins 35
The inverted tree 37

viii Contents

    Christianity vindicated 40
    The Hindu–Catholic fire-eater 44
    Conclusion 46

4 Worldviews in encounter   48

    Early Indian science 49
    External influences 52
    Classical Indian science 54
    Science under Islam 56
    European philosophy and science 59
    Darwin, evolution and progress 60
    The nature of the universe 64
    Conclusion 66

5 Relativity and beyond   68

    The road to relativity 68
    Quantum theory 70
    General relativity 71
    Bose–Einstein statistics 73
    The Uncertainty Principle 74
    Towards a unified field 77
    A road less travelled 78
    State of the art 79
    Conclusion 82

6 Indian science comes of age   84

    Science becomes institutionalized 84
    The rediscovery of ancient science 86
    Eminent scientists 88
    Conclusion 100

7 An investigation into the beliefs of Indian scientists   102

    Framing the hypotheses 103
    Previous research 105
    Results of the investigation 106
    Religious affiliation 108
    The relationship between science and religion 109
    Investigation of the hypotheses 114
    Comparison between Hindus and Christians 118

The influence of science 119
Issues of historical significance 122
Discussion of the results and conclusions of investigation 127

8 How clear is reason's stream? 130

Tagore's major themes 130
The God of humanity 132
Einstein's religion 135
Spinoza's God 137
Meeting of minds 138
Dialogue in context 140
Another view of science 141
Contemporary perspectives 143
Conclusion 149

9 Looking to the future 151

Avoiding the pitfalls 155
Hidden unity 156

Appendix A: The nature of reality	161
Appendix B: Investigation questionnaire (Chapter 7)	165
Select glossary of terms	167
Notes	169
Select bibliography	180
Index	183

Illustrations

Figure

5.1	The Standard Model of elementary particles	81

Tables

7.1	Courses and institutions of the respondents	107
7.2	Religious affinity of high school attended	108
7.3	A comparison between personal and parents' religious affiliation	108
7.4	Whether conflict perceived between religion and science	109
7.5	Influence of degree course	109
7.6	Direction of change in beliefs due to influence of degree course or scientific research	110
7.7	A comparison of responses to specific areas of possible conflict between science and religion	111
7.8	Importance attached to religion	111
7.9	Extent of agreement with Einstein's statement	112
7.10	The importance of different authorities in making ethical decisions	112
7.11	Frequency of attendance at places of worship	113
7.12	Interpretation of God	113
7.13	Distribution of responses to reincarnation as a function of the importance attached to religion	115
7.14	Chi Square and the levels of significance for perceived conflict between religion and science and influence of degree course	116
7.15	Chi Square and the corresponding levels of significance for evolution and reincarnation	116
7.16	Importance of religion analysed in terms of attendance at places of worship	117
7.17	Values of Chi Square for Hindus and Christians according to the degree of importance attached to religion for a selection of responses to questions 4 and 5	119

Acknowledgements

This book was first suggested by Professor R. Ninian Smart, and I have therefore dedicated it to him. His idea began to come to fruition in February 2005 when I was asked to give the Teape seminars at the University of Cambridge. I am therefore grateful to the Trustees of the Teape Foundation and the Cambridge–Delhi Christian Partnership for this honour, and also for inviting me to give the main Teape lectures in India in 2008.

I also wish to express my gratitude to the following: John W. Bowker, Prem Sagar Dwivedi, Julius J. Lipner, Daniel O'Connor and members of the Society of Ordained Scientists for their assistance.

I am grateful to Andrew Robinson for permission to use the photograph of Einstein with Tagore, published in *Einstein: A Hundred Years of Relativity*, 2005.

Rosemary Smith painstakingly typed and gave stylistic advice on the manuscript, and I am enormously grateful to her for that.

David L. Gosling
Clare Hall, University of Cambridge

1 Introduction

Tradition is generally understood to include a substantial component of religion, which many people believe to be incompatible with science. It is therefore not surprising that there is a large body of literature which attempts to reconcile science with tradition, much of it by scientists who are religious believers. But these publications are predominantly concerned with the situation in the West, and very little research has been done on science within the context of the Indian tradition, of which the Hindu element is the most unfamiliar to European Christians, and also very complex. What follows attempts to correct this imbalance and open up new avenues for understanding the relationships between science and religious belief in an inter-faith context.

Science in India

Early in 2005 the *New Scientist* devoted a special edition to 'India, the next knowledge superpower'.[1] While the first indications had appeared eight years previously with claims in the media that young Indians were 'stealing' information technology (IT) jobs from wealthy nations, the trickle has become a flood. Between 2000 and 2005 alone, more than a hundred science-based and high-tech companies have set up research laboratories in India, recruiting their personnel from India's 250 universities and prestigious technology institutes – a pool of just under 3.5 million science graduates in all. During the same five years the contribution of the IT industry to India's economy in terms of Gross Domestic Product has risen from 1.5 per cent to just over 3 per cent.

India's pharmaceutical industry has undergone a comparable revolution. Thanks to distinctive patenting laws which allow patents to be taken out on the processes whereby medicines are made, but not on the actual medicines, cheap generic versions of some of the West's most important drugs have been produced at a fraction of the price for sale in India and other developing countries. Thus life-saving therapies for AIDS which cost US$10,000 a year in the West have been made available in Africa for less than US$200. New international patenting rules have now made such undercutting of Western products more difficult, but the expertise gained has led to the creation of approximately 9000 Indian pharmaceutical companies. Herbal therapies, a major part of India's traditional ayurvedic medical system, are being tested in the most modern laboratories, in some cases improved, and marketed in India and for an increasing global market.

India's president, Dr A. P. J. Abdul Kalam, is a distinguished representative of two scientific fields in which the country excels, space science and nuclear power. It was Indira Gandhi – India's longest-serving prime minister – who maintained that a successful space programme is not only important for science, but essential for national development. Thus India's six remote-sensing satellites observe rivers and coasts to enable rural and coastal communities to anticipate geographical and climatic changes. Seven communication satellites beam television and extend education and healthcare to the rural poor. The tsunami which hit India's eastern coasts late in December 2004 deposited seven metres of seawater on top of the foundations of the country's first commercial fast breeder reactor – an unanticipated setback to a nuclear power programme stretching back half a century. India is short of oil and natural gas; indigenous coal is available, but a major pollutant; and hydro-electricity and most of the renewables are subject to geographical limitations. The argument for nuclear power is therefore strong though some are opposed to it.

Medical research and treatment, which incorporates spin-off techniques from nuclear power research, stands at a high degree of capability. In Hyderabad, for example, the L. V. Prasad Eye Institute is pioneering a stem-cell cure for levels of eye damage extending beyond the cornea and limbus to the conjunctiva. In spite of the high degree of sophistication of this research, the charitable status of the institute means that the consequent therapies can be made available at minimal cost. Other major areas of science and technology in which India excels include computer engineering, aeronautics, radio astronomy and agricultural research on genetically modified (GM) crops. GM crops developed in India are less open to failure and exploitation than those imported from abroad.

While the benefits of progress in science and technology have been assimilated primarily among the burgeoning middle classes – variously estimated at between 150 million and 250 million out of a total population of just over a billion, poorer people have been quick to utilize high-tech products. Fishermen now price their catches using mobile phones before landing, autorickshaw drivers take orders by phone (the quality of their lives having been improved enormously by a switchover in several cities from diesel fuel to compressed natural gas), and solar panels are being installed in remote areas to produce electricity (one such device is in use at a Vaishnavite (Hindu) monastery on Majuli Island in Assam, another at a mosque in Srinagar, Kashmir).

Problems remain: there is a gap between the academic world and industry; water, road and rail projects are often undertaken half-heartedly; overemphasis on IT and management has drained some of the brightest students away from science; and research remains underfunded. But if these and other mismatches can be negotiated successfully, it is estimated by leading investment banks in North America that by 2050 India will possess the third largest economy in the world, after the USA and China. And science will have played a large part in the transformation.

Science and the environment

No account of science in society is complete without an acknowledgement of its impact on the natural environment and its potential for improving it. In India this became a significant issue in the 1970s, originating at the political level with Mrs Gandhi's contribution to the United Nations (UN) Conference on the Human Environment in Stockholm in 1972. In her speech she described the wanton destruction of forests and biodiversity, matching this with the importance of meeting the needs of the poor:

> The environmental problems of developing countries are not the side effects of excessive industrialisation but reflect the inadequacy of development.... [But] how can we speak to those who live in villages and in slums about keeping the oceans, the rivers and the air clean when their own lives are contaminated at the source? The environment cannot be improved in conditions of poverty.[2]

The importance of holding together developmental and environmental needs was acknowledged by subsequent UN conferences, culminating in the Earth Summit held at Rio de Janeiro in 1992, and its successor, which took place in Johannesburg ten years later. The role of science is to address these twin areas of need as far as possible together. Thus, for example, it is no longer responsible science to build a large dam on a river to provide electricity for the manufacture of some shiny new technology, if in the process communities above and below the dam are flooded and deprived of their livelihood. This is an extremely controversial issue in India.

Climate change came to the forefront of international concern in the 1980s, and led to the World Conference on the Changing Atmosphere in Toronto in 1988. The Indian argument from the start was that since the industrial countries were largely accountable for the emissions responsible for global warming, they should make the greatest effort to reduce them. But for India this issue has constantly been bedevilled by misleading data and statistics. Thus the Washington-based World Resources Institute put out a pre-Summit report claiming that greenhouse gas emissions from developing countries are far more extensive than is the case. The Tata Energy Research Institute and the Centre for Science and the Environment, both based in Delhi, successfully challenged their figures. One hundred and fifty-four countries, including India, but not the USA, finally signed the climate change convention.

Nowadays misinformation about global warming has become more sophisticated. Thus we are frequently told by Western news agencies that India is the sixth largest global emitter of carbon dioxide. Maybe, but then the single state of Uttar Pradesh, taken out of India, would be the seventh largest nation in the world. More pertinently, the *per capita* production of carbon dioxide is 10 per cent of that of an average American.[3] A lot is at stake in the global warming blame game, and vested interests are ruthless in bending science to suit their wishes.

During the 1970s and 1980s a considerable number of non-governmental environmental groups appeared in most Indian states. They are essentially groups of concerned citizens worried about deforestation, waterway and other types of pollution, basic healthcare, women's education, resource depletion and the essential needs of predominantly rural communities. It is remarkable that one hundred registered as non-governmental organizations (NGOs) at the Earth Summit.[4] A number of these couched their aims in cultural or religious (largely Gandhian) terms, as did the Indian Government's official submission, which included quotations from the Upanishads and reference to sacred groves, Ashoka's pillar edicts, the Chipko and Appiko movements, and Gandhian ethics.[5]

Unlike their counterparts in the West, which have progressively adopted an anti-science ideology, members of Indian environmental organizations are proud of their country's scientific achievements. This does not mean that they are totally uncritical of GM crops, nuclear power and the other *bêtes noires* of Western environmentalists. But their responses to new technologies are more in the form of questions and calls for more or improved testing of products than outright opposition. Sunita Narain, a research chemist, directs the Society for Environmental Communications (Centre for Science and the Environment), and edits *Down To Earth*, a penetrating and lively fortnightly magazine mainly devoted to Indian environmental issues, but including perceptive critiques of international topics. It is interesting, for example – and pertinent to the overall concerns of this book – to see what such an influential Hindu scientist has to say about the current Western preoccupation with creationism and 'Intelligent Design':

> The question of the epistemological status of religion and science is best showcased in the creationism debate. In several countries of Europe and the US it has been argued that students should be given the choice of studying creationist, Biblical theories as an alternative to scientific theories about the creation of the universe. The idea is sheer bunkum.
>
> Creationist theories are not based on any form of verifiable evidence, and do not follow the principles of testability, the very hinges of science. Doubtless, people are free to believe that [God] created the world. But that belief cannot be elevated to the status of science and taught in a science curriculum...
>
> Religion, a philosopher famously said, is the heart of a heartless world, the sigh of the oppressed creature. Thus far the going is good. But what it cannot do is set the conditions under which the pursuit of knowledge of the physical world or scientific innovation must happen.[6]

It is refreshing to read such a forthright condemnation of Western obscurantism dressed as science by a leading Hindu science editor!

Regretfully, in the chapters that follow, we shall not be able to quote from many Indian women scientists. The reasons for this omission are complex. According to *Frontline*, a leading weekly news digest published by the *Hindu* group of newspapers in Chennai,

> There is an overwhelming body of empirical and qualitative evidence to suggest that a strong gender bias pervades institutions of science in India....

Although there is no explicit discrimination against women in enrolment and recruitment at the college, university or faculty levels, attitudinal biases against women and unsupportive institutional structures have...operated as powerful forces against talented women realising their full potential in...careers in science.[7]

India is by no means alone in this respect. Out of 776 Nobel Prizes awarded in the fields of physics, chemistry and medicine (and from 1969 economics), only 34 have gone to women!

Society and tradition

The social dimension is often the most appropriate point of entry into India's religious life, and this is as true of young Hindu scientists as of other sections of the community. Following a debate about reincarnation organized by the Hindu Society of the University of Cambridge, I asked a postgraduate student researching micro-organisms where he came from. On learning that he had studied in the catchment area for the Indian Institute of Science in Bangalore, where I once worked, I enquired about his social origins. 'Are you asking me my *caste*?' he retorted indignantly. I changed the subject, and we happily discussed the implications of science for reincarnation, which he believes in. Later he told me that his family are Iyer brahmins, but that although the Iyers are Śaivites (i.e. believers in Śiva), he prays to God as Krishna (i.e. an incarnation of Vishnu), who is his friend.

The initial sensitivity of this young scientist about his caste contrasted with that of research scientists I met at the Indian Institute of Science, who readily identified themselves as Iyers, Smārthas, Iyengars, Saraswats, Madhvas, non-brahmin Śaivites and Vaishnavites, Muslims, Sikhs and Christians. These groupings were also evident from where people sat at meal-times. Community, caste and family are the traditional determinants of Hindu identity, and since Hindus account for 81.3 per cent of India's population, the remainder are usually classified according to their community. The corresponding figures are as follows: Muslims, 13.4 per cent; Christians, 2.3 per cent; Sikhs, 1.9 per cent; Buddhists, 0.8 per cent; Jains, 0.4 per cent; plus some others, such as the Parsis. These are the communities which constitute Indian society, and there is enormous variety among and within them.

The social dimension of a tradition represents only one aspect of its existence, but in most of Asia it is foundational. Socio-religious rituals and personal experience are also significant, whereas beliefs about God (or gods, that is, 'celestials') and ethical behaviour are more important in Western societies. By 'Indian tradition' we mean – following Ninian Smart – all the social, ritual, experiential, doctrinal and ethical dimensions of the various communities which make up the sub-continent.[8] An entire library could not cover such a vast area!

Sometimes the various manifestations of 'tradition' are best described as religious, and sometimes in terms of culture. There is no simple way of making a distinction. Some people regard all religions as divisive and superstitious, but respect culture. Others consider culture to be subversive (e.g. the German Nazi who is alleged to have said, 'Whenever I hear the word culture, I reach for my gun'). According to M. M. Thomas, one of India's most distinguished Christian thinkers:

> Culture is an embodiment of values in forms capable of influencing imagination and affections, and therefore of motivating behaviour, a dimension between religious truth and social institutions. It shapes the distinctive character, the selfhood, of a people.[9]

Even science has cultural overtones, as the following definition by Robert Merton, an American sociologist, makes clear:

> Science is a deceptively inclusive word which refers to a variety of distinct though inter-related terms. It is commonly used to denote a set of characteristic methods by means of which knowledge is certified; a stock of accumulated knowledge stemming from the application of these methods; a set of cultural values and mores governing the activities termed scientific or any combination of the foregoing.[10]

Science is universal, and although some scientists may undertake their research in the belief that its progress reflects their religious outlook, there can be no 'Hindu' or 'Christian' science, as such. Science is often wrong, but it is at least wrong about *something*, and no quarter can be given to those who try to relativize or deconstruct it. The following statement by the seventeenth-century thinker Isaac Pennington illustrates how science proceeds:

> All truth is a shadow except the last. But every truth is substance in its own place, though it be but shadow in another place. And the shadow is a true shadow, as the substance is a true substance.[11]

Religion is generally understood to mean the acknowledgement of a world-transcending reality (or realities), personal or not, in which human fulfilment may be achieved. Some world-transcending realities are described in personal terms (e.g. Allah or Yahweh), others, such as the Buddhist goal of *nirvāṇa* (Sanskrit) or *nibbāna* (Pali), are not. The former may be said to have a 'divine', and the latter a 'trans-divine' focus. The various facets of India's multifarious religious traditions will be considered, but it will be useful to note here certain aspects of the Hindu tradition. 'Hinduism' – a term which we shall tend to avoid on account of its monolithic overtones – is a family of culturally similar traditions, their culture based geographically on the sub-Punjab area of Sind (now in Pakistan), where a river described in the Vedic scriptures as the Sindhu (probably the Indus) once flowed.[12] Thus culture and geography are sometimes more important determinants of the Hindu tradition than religious doctrines. The Hindu tradition is a unity-in-diversity.

Science and religion

The Indian tradition originally made no principled distinction between 'natural' and 'human' sciences, or between science and religion. God (or gods or 'the One'), humanity and nature flowed into one another to such an extent that science and religion – as considered today – were part and parcel of a single body of knowledge, the Veda (which means 'knowledge').

In the West, for a variety of reasons, science and religion moved apart, and there is now an industry of literature dedicated to explaining why this happened and what should be done to close the gap. Most of this is geared to reconciling Protestant and Roman Catholic doctrines and aspects of science which, superficially at least, appear to challenge them: cosmology, evolution, and so on. These debates are of concern here only insofar as they impinge upon India.

Why should science and religion be considered together at all? Granted that science seems new and shiny, and appeals to the young, whereas religion often appears backward-looking and unattractive, it is not difficult to understand why educated believers try to reconcile the tenets of their faith and science. The common ground between science and religion is that they claim to depict reality – both attempt to point beyond, or transcend, the familiar matrices of our daily lives. Religion directs our attention towards God, the One, *Brahman*, or *nirvāṇa* (which may be divine or trans-divine), and usually also provides us with consequent patterns of appropriate behaviour.

There are important differences between the use of reality-depicting models in science and in religion. According to Janet M. Soskice,

> There are... models in both scientific and religious language which are used to speak of things beyond our immediate experience, but there are these differences: scientific models are used to explicate and schematize a theory, they are projective, that is, they suggest new possibilities of understanding and most importantly they do so with the aim of accounting for causal mechanisms. Religious models, on the other hand, do not embody theories or explanations but serve to evoke a response from the listener, and to call to... mind or present... an issue with forcefulness. Consequently, while the models of science are abstract and objectifying, the models of religious language are personal; while the models of science are used with precision, their entailments tightly constrained by the formal theory they illustrate, the models of religious language are... non-assertive, and always 'qualified'. And, finally, whereas in scientific thought the subservience of the model to the theory means that the model is a useful but dispensable heuristic, in religion there can be no comparable retreat into pure theory, for in religion the models are all that we have, trapped... within... 'a wheel of images'.[13]

Put slightly differently, the models of science are causally explanatory and structural; they are aids to theorizing which can be replaced by a theory or set of equations. Religious models are evocative and may carry ethical implications; they cannot be substituted by equations, and can only be replaced by equally empirically unverifiable models. There is therefore little to be gained by trying to use scientific models to validate or endorse religious claims. Science may overreach itself (and often does – we use the term 'scientism' to describe this tendency), but religion does not need to compete with science to deserve to be taken seriously.

Literature seeking to relate science and religion is extensive, and most of it is concerned with the Abrahamic religions, especially Christianity in the West. John Hedley Brooke's *Science and Religion: Some Historical Perspectives* contains references to Islam and Buddhism, but does not mention the Hindu tradition.[14] Christopher Southgate's *God, Humanity and the Cosmos* brackets the entire Hindu family of traditions under the common head of 'panentheism'.[15] Keith Ward's choice of Aurobindo in his otherwise excellent *Religion and Creation* is hardly representative of the major schools of Hindu thought.[16] There is a small but growing body of literature in India which attempts to relate science and religion.

Einstein meets Tagore

Albert Einstein and Rabindranath Tagore are legendary figures whose reputations endure into the twenty-first century. Einstein's centenary in 2005 was a global event which served to demonstrate how little, at its deepest level, science has moved beyond the issues with which he struggled. Tagore's universal humanism, passionately expressed in his poems and publications, is just as enduring – so much so, for example, that the Marxist chief minister of West Bengal can recite 500 of his songs from memory.[17] Einstein and Tagore met for the first time in 1926 in Germany. Their first conversation about the nature of reality took place on 14 July 1930 at Einstein's home at Kaputh, near Potsdam. We include it as Appendix A. Their second and third meetings were at Kaputh; the fourth occurred in mid-December in New York. Appendix A also includes the recorded part of the second conversation.

Neither Einstein nor Tagore were particularly happy about their recorded conversations, and modified them prior to publication. This may in part be due to the fact that each of them would have preferred to communicate directly in his mother tongue – German or Bengali. As it was, Tagore spoke in English and Einstein spoke German, which was translated. But for all that their minds never seemed to truly meet, they shared a deep mutual respect. Einstein alluded to Tagore affectionately as 'rabbi' (teacher), and Tagore turned down the offer of a doctorate from Berlin University as a protest against the Nazi treatment of Einstein. In what follows mention will be made of these conversations, the first of which has also been used to stimulate discussion among several Indian scientists. But the main thrust will be to delineate the progressive encounter between Western science and educated India during the last two centuries, and the development of science in Indian society.

Methodology

The following chapters contain both historical and sociological material. The historical material charts the incorporation of western science into India from the early nineteenth century onwards, following the decision to promote higher education in the English language in 1835 until approximately a century later when Einstein and Tagore encountered one another. This period is crucial for any understanding of the Indian tradition today because it covers the growth of reform movements which transformed the beliefs of educated Indians – Hindus, Muslims and, later, Christians. It is also vital from the point of view of science because it includes the contributions of a group of remarkably gifted scientists, mostly, but not exclusively, from Bengal, who shaped their research in accordance with their religious and philosophical beliefs.

The sociological material is based on an investigation conducted by the author, supplemented by interviews obtained during the last three years. It was carried out at a number of scientific institutions in four centres, Delhi, Bangalore, Kottayam and Madurai, and has been partially published in India.[18] The rationale for combining historical and sociological material has been given in a previous volume.[19] Endorsement was given to the views of Louis Dumont and David Pocock to the effect that the 'little tradition' of contemporary India and the 'great tradition' of the classical religious texts are one and the same, and that their unity is based on a reiterated relationship which can be traced in different areas of life.[20] This school of thought has been most convincingly articulated by Stanley J. Tambiah, who was until recently the head of anthropology at Harvard University.

Tambiah criticizes the notion of two levels in religion – the higher literary and the lower popular – on the grounds that it is 'in some respects static and profoundly *ahistorical*'. In fact, most religions are constantly changing both their beliefs and structures, and the classical texts 'range over long periods of time and show shifts in principles and ideas'.[21]

> In the study of religion...I would make a distinction between *historical* religion and *contemporary* religion, without treating them as exclusive levels. Historical [religion] would comprise not only the range of religious texts written in the past, but also the changes in the institutional form of [religion] over the ages. Contemporary religion would simply mean the religion as it is practised today and should include those texts written in the past that persist today, and are integral parts of the ongoing religion. Thus, if the question of the relation between historical and contemporary religion interests us, we should look for two kinds of links, namely, *continuities* and *transformations*.[22]

Tambiah's studies of Thailand, which received little exposure to the West compared to India, provide ready examples of such continuities and transformations, and we identified some in our previous study of ecology and religion in South Asia. One such transformative link is the manner in which the Hindu scientist, Jagadish Chandra Bose, directed his work consistent with his belief in *advaita* Vedānta into researching the possibility of pain in plants.

The anthropological material is based on face-to-face interviews. These have been edited to form a continuous narrative which omits the questions, which are usually self-evident. The following example is based on discussions with

Dr S. V. Eswaran, head of the chemistry department at St Stephen's College, Delhi University, who was interviewed in March 2004:

> My scientific work reaffirms my faith in higher powers. There is so much more to know and therefore there must be something infinite. Physics points towards a unified field theory. Chemistry studies atoms and molecules in which precision and order are inbuilt. A force or God keeps it all together. There is a larger plan in nature – it is not haphazard. This suggests an organizer. Science proceeds in quantum leaps and this is not a matter of chance.
>
> I think in terms of spirituality rather than religion as an influence on my beliefs. There is a natural transition between science and spirituality and there is no conflict. Religious rituals are merely simplifications.
>
> Among the brahmins of the top Indian scientists I was a mere college lecturer – a *śūdra*. After thirty-five years of my scientific work, I thought that as I reached retirement my career would come to an end. But God answered my prayers, allowing me to do more and more. A spiritual force helped me, and my new openings in research and teaching I owe to my spiritual experience.
>
> I myself am a brahmin and I have been guided by Satya Sai Baba, who lives in Puttaparthy near Bangalore. God's grace alone has ensured my being scientifically highly productive in the last decade. In my formative years, I was guided by Swami Ranganathananda, who heads the Ramakrishna Math and Mission at Belur near Kolkata. These are saintly people who illuminate the lives of others. I have also been inspired by Professor T. R. Seshadri, organic and natural products chemist, who was also a Fellow of the Royal Society, London, and by Professor A. Butenandt, past President of the Max-Planck Society in Germany. His colleague Professor W. Schäfer once called me and complained that I was not visiting the museums and art galleries in Munich to learn more about German life and culture, to which I replied: 'Sir, for me you are yourself a living museum and working with you in your laboratory tells me more about Germany than visiting any museum or gallery.'
>
> I try to live such that every act is a prayer. I believe that reincarnation is as natural as the transmutation of chemical elements into one another. I read the Bhagavadgītā and I can chant from the Vedas – I do it in public. Words have great power and can generate calmness. We must do everything possible to remove the self (ego) and engage in *niṣkāma* karma (selfless work).

Conclusion

The chapters that follow begin with a broad narrative of the events and personalities which shaped science and religion in modern India (Chapter 2). Chapters 3 and 4 explore in more detail some of the important issues arising from this narrative. Among the nineteenth-century trends which Indian scientists and others saw as having religious significance was the tendency of discrete sciences to converge under common theories; Chapter 5 takes up this theme in relation to Einstein's work and beyond, and paves the way for the discussions which occurred between him and Tagore in 1930. It also offers an example of an alternative view of atomic science.

Chapter 6 charts the manner in which science became institutionalized following the decision to use the English language as the medium of instruction for higher education, and the contributions of several outstanding Indian scientists, some of whom collaborated with Einstein or shared his vision of a unified field theory. Chapter 7 summarizes an investigation by the author into the religious beliefs of Indian scientists at four major centres. Chapter 8 offers an account of Tagore's attitudes to science and religion and Einstein's understanding of religion, and discusses their two recorded conversations. Chapter 9 draws together the various threads, and suggests that what has hitherto been a narrow and essentially Western encounter between science and religion needs to be broadened – to their mutual benefit – by taking on board the insights and experiences of scientists and religious thinkers – Hindus, Muslims and Christians – in India.

As a general rule technical terms have been italicized unless they are well known (e.g. karma) or have been Anglicized (e.g. brahmin, Shankara). Diacritical marks are used with Sanskrit but not Hindi words. Quotes are given in their original form. There is a select glossary of frequently occurring technical terms at the end.

2 Science in India's intellectual renaissance

An appreciation of India's intellectual renaissance, which occurred during the second half of the nineteenth century, flowing into the twentieth, is vital in order to understand modern India. It was within this renaissance that Western science took root and flourished to a remarkable extent in a group of brilliant scientists some of whom collaborated with Einstein in his groundbreaking work. The conversations which occurred between Einstein and Tagore in 1930 mark essentially the end of this fertile period, after which the energies of educated India were directed increasingly into the struggle for independence, while Europe was plunged into a terrible war.

We shall consider the main features of the nineteenth-century renaissance, and the role of science within it, noting in particular the significance of the decision to introduce higher education via the medium of the English language. Our primary concern will be with the narrative of events as they occurred, and the ideas and beliefs of those who shaped them. A later chapter will consider in more detail the relative states of science in India and Europe prior to and during the nineteenth century, and the distinctive features of the religious and philosophical beliefs which shaped India's response to the West. Tagore is discussed briefly at the end of this chapter, and in more detail later. Of necessity we shall be dealing with English-language sources in an attempt to give a broad sweep of the period under consideration. For a more detailed analysis there are excellent studies by Deepak Kumar, Dhruv Raina, Gyan Prakash and others.[1]

Macaulay's Minute

The decision to introduce a system of English-medium education into Indian higher education in 1835 played a decisive role in the development of science in India. The so-called Anglicists, who supported the decision, were a mixed bag of utilitarians, who wanted scientifically literate civil servants to carry out the work of government, missionaries, who expected Western rationalism to precipitate the collapse of Hindu superstition (as they saw it), leading Hindu intellectuals, such as Ram Mohan Roy (1772–1833), who argued that the 'Baconian philosophy' might one day do for India what it had achieved for the West, and others.

Science in India's intellectual renaissance 13

In 1823 Roy wrote to Lord Amherst, Governor-General in Council, as follows:

> I beg your Lordship will be pleased to compare the state of science and literature in Europe before the time of Lord Bacon with the progress of knowledge made since he wrote. If it had been intended to keep the British nation in ignorance of real knowledge, the Baconian philosophy would not have been allowed to displace the system of the schoolmen which was the best calculated to perpetuate ignorance. In the same manner the Sanscrit system of education would be the best calculated to keep this country in darkness, if such had been the policy of the British legislature. But as the improvement of the native population is the object of the government, it will consequently promote a more liberal and enlightened system of instruction, embracing mathematics, natural philosophy, chemistry, anatomy, with other useful sciences, which may be accomplished with the sums proposed.[2]

Against such arguments the 'Orientalists', led by distinguished Sanskrit and Persian scholars, argued that India's rich heritage of learning – which, as we shall see in a later chapter, included much science and philosophy – should be disseminated in these traditional languages.

In 1835 the recently appointed president of the Committee on Public Instruction, Thomas Babington Macaulay (1800–59), carried the day for the Anglicists. The following argument was central to his Minute:

> Whether we look at the intrinsic value of our literature, or at the particular situation of this country, we shall see the strongest reason to think that, of all foreign tongues, the English tongue is that which would be the most useful to our native subjects.
>
> The question now before us is simply whether, when it is in our power to teach this language, we shall teach languages in which, by universal confession, there are no books on any subject which deserve to be compared to our own; whether, when we can teach European science, we shall teach systems which, by universal confession, whenever they differ from those of Europe, differ for the worse; and whether, when we can patronize sound philosophy and true history, we shall countenance, at the public expense, medical doctrines which would disgrace an English farrier, astronomy which would move laughter in girls at an English boarding school, history abounding with kings thirty feet high and reigns thirty thousand years long, and geography, made up of seas of treacle and seas of butter.[3]

Lord Macaulay's contemptuous dismissal of popular beliefs based on the Purāṇas did considerably less than justice to the Hindu tradition as a whole, much of which was in process of being made available in English for the first time by English scholars. Although many ordinary Bengalis may have subscribed to the views which he ridiculed, there was an emerging body of upper-caste Bengalis often referred to as the *bhadralok* (gentility) who were to prove fertile soil for a much more sophisticated and intellectually credible understanding.

It is important to recognize that although not all sections of the British Raj shared Lord Macaulay's prejudices, it was none the less a deliberate and continuing element of colonial policy to use science as an indispensable tool of colonial domination. According to Peter Bowler,

> The proclamation of Queen Victoria as Empress of India in 1877 heralded the emergence of the new imperialism in which European governments set out to conquer and control territory by exploiting the latest technology.... Whether the expansionist interests were commercial or military, however, science was seen as an indispensable tool.... [Thus] Kew Gardens became the hub of the British empire's efforts to replace indigenous species with imported ones of greater commercial value.[4]

Such foolish and shortsighted policies did immense damage to delicately balanced forest ecosystems.[5]

The secularization of belief

Indian responses to influences from the West have been variously classified. Bhikhu Parekh describes them in terms of modernism, critical modernism and critical traditionalism.[6] For the modernists society needed to be restructured along European lines. The critical modernists pleaded for a creative synthesis of the two civilizations. The critical traditionalists wanted to mobilize their indigenous resources, borrowing from Europe where necessary. These three categories are not mutually exclusive:

> The traditionalists, the modernists and the critical modernists were all in their own different ways convinced that civilisations could be compared and assessed on the basis of some universal criteria. The critical traditionalists, who included the later Bankim, Vivekananda, B. C. Pal, and Aurobindo, rejected the assumption. For them, a civilisation was an organic whole and could not be judged in terms of criteria derived from outside it. All such criteria were themselves ultimately derived from another civilisation and thus lacked universality. Furthermore, values and institutions were an integral part of the way of life of a specific community and could not be judged independently of its capacities, habits, dispositions and deepest instincts. What was good for others might not be good for it.[7]

Within these overlapping categories of response to the West, Parekh sees science as a common denominator:

> Modern science became extremely popular and almost every leader turned to 'scientific research' and 'scientific method' to generate 'scientific ethics' and

'scientific principles of society'. Bengal was for a while full of Comte and found in his positivism a method of discovering new truths. Ranade advocated 'Bacon's method', Gokhale thought that J. S. Mill's 'method of empiricism' alone was reliable, and Aurobindo turned to a combination of Darwin and Einstein.[8]

Parekh's categories will be noted from time to time. However, for practical purposes – and also to simplify matters – we shall classify the various reform movements within the Hindu tradition and to a lesser extent in Islam and Christianity in terms of responses to secularization. This we define, following the anthropologist M. N. Srinivas, as the process whereby areas of life once governed by tradition cease to be such.[9] The subsequent responses may be classified in terms of the reassertion, rejection or adaptation of tradition (with possible overlap). These responses will be illustrated with particular reference to the manner in which science was utilized to shape them. The subsequent blossoming of reformed versions of the Hindu tradition and, to a lesser extent, Islam during the second half of the nineteenth century and beyond was many-faceted – social, religious, cultural and scientific.

In the introductory chapter it was noted that the Hindu tradition is essentially a family, a unity-in-diversity. But under the onslaught of Western philosophies and religion, many educated Hindus began to focus their beliefs on what seemed to be the most resilient system (*darśana*) of philosophical and religious thought, namely *advaita* Vedānta, as set out by Shankara (600–700 CE). Parekh notes this tendency as follows:

> Most of [the reformers']...ideas were grounded in *advaita*, which was therefore taken to be the 'characteristic world view' of India and ideally suited to what Ram Mohun Roy called the 'basis of Indian unity'.[10]

Ram Mohan Roy was, in Parekh's terms, a critical modernist who sought to combine Vedānta with modern scientific culture. He started the Vedānta College in 1825, followed by what became the Brahmo Samaj. This was the most potent vehicle of secularization which produced irreversible social and religious reforms throughout the nineteenth and early twentieth centuries. Rabindranath Tagore was a leading representative of later Brahmoism.

Roy was born in rural Bengal in a home where Vaishnavite *bhakti* (devotionalism) prevailed. His schooling at a Muslim *madrasa* would have deepened his belief in monotheism prior to his arrival in Calcutta where he began a systematic campaign to reform Hindu beliefs and customs. A major component of Roy's reforms was his reinterpretation of *dharma* as a rationalist ideal, based on egalitarian religious ethics. He believed that everybody can have access to God, and that all people can potentially achieve the same destiny. He even went so far as to reject reincarnation.[11] His fundamental framework of Hindu belief was shaped by Shankara's *advaita* Vedānta, though this came to the fore mainly when he was speaking and writing in Bengali ('Brahmo' is Bengali for *Brahman*). At other times, as when in conversation with

Christian missionaries, for example, his outlook was much more dualistic. He was fluent in Persian, and wrote treatises on monotheism in Arabic.

Roy was an immensely articulate propagandist for his country and the reformed Hindu beliefs which he espoused. He persuaded at least one orthodox Christian missionary to give up his faith and become a Unitarian. On visits to London he astonished his audiences with his knowledge of both the Sanskrit and Persian streams of Indian tradition and of Western – including French and German – philosophy and religion. His *Precepts of Jesus* displayed a greater familiarity with New Testament Greek than that of most missionaries and the London gentility who flocked to hear him. On his final visit the English climate proved too much for him; he died, and is buried in a Bristol cemetery.

Educated Hindus had to accommodate not only to new ideas, but also to traditionally prohibited practices, such as human dissection. The Calcutta Medical College, founded in the year in which Lord Macaulay published his Minute on education, was within a decade producing students who won the highest academic honours in England. The college's official history records that a student called Babu Raj Krishna was the first 'to plunge the scalpel into the dead human body'.[12]

The first signs of a response among Muslims to Western science were at the royal court in Lucknow (as recorded by Bishop Heber) and at the Delhi College in the 1830s. By 1844 Professor Yesudas Ramchandra was lecturing in science with a pioneering enthusiasm which left its mark not only on his students, but on much of the city. In his *Memoirs* he records:

> The doctrines of the ancient philosophy taught through the medium of Arabic were cast into the shade before the more reasonable and experimental theories of modern science... But the learned Maulwis who lived in the city... did not like this innovation.[13]

It was not only in the colleges that science was welcomed. Several popular journals such as the *Samachar Darpan*, published from Serampore, the *Sambad Prabhakar*, a widely read if somewhat conservative daily founded in 1839, and the *Tattvabodhini Patrika*, established by Debendranath Tagore (Rabindranath's father) in 1853, all contained positive articles and editorials about science: 'No useful purpose is served by teaching arts and literature. The work of Kalidas, Shakespeare and others may provide literary pleasure but there will be no real progress without scientific instruction.... No country can progress without the advance of technology'[14] – thundered the *Sambad Prabhakar*!

The missionaries had initially welcomed science: 'In the English language with its true literature and science, we have an engine by which, if rightly wielded, the most towering superstitions and idolatries of the East might be levelled.'[15] Such was the opinion of the Scottish missionary, Alexander Duff, who collaborated with Ram Mohan Roy in setting up a school. Duff never tried to introduce science into religion classes; others did, but with mixed results:

> We who are engaged in this... work cannot take the position of Paul and say 'I determined not to know anything among you save Jesus Christ and him

crucified.' There would be more truth in saying 'I am constrained to do little among you save Fowler's *Logic* and Todhunter's *Algebra*.'[16]

By the first Decennial Missionary Conference in 1872 it was reluctantly acknowledged that the strategy of using science to level religious superstitions had failed: 'The higher education of heathen minds sets them the more against us; they use their education as a club to break our heads', lamented one participant![17] But although the consequences of introducing the English language and science into higher education had disappointed some, there were other areas in which the results were extremely fruitful. The Reformers, so-called, both Hindu and Muslim, and a remarkable generation of Indian scientists evolved from the ferment which arose during the first half of the nineteenth century.

The Reformers

The Brahmo Samaj was formally constituted in 1843 under Debendranath Tagore (1817–1905). Debendranath set his thinking firmly within the Upanishadic tradition, and moved away from the monotheism of Islam and Christianity. He was less influenced by science and technology than Roy, conceiving of *advaita* Vedānta as a counter to Western dualism. God could be known primarily through 'intuition'. This notion was also invoked by the Bengali chemist, P. C. Roy, to describe the act of scientific discovery.

Keshub Chunder Sen (1838–84) was the third leader of the Brahmo Samaj. His acceptance of science was totally uncritical:

> Science will be your religion... above the Vedas, above the Bible. Astronomy, geology, botany and chemistry, anatomy and physiology are the living scriptures of the God of Nature.[18]

Keshub's beliefs were strongly influenced by evolutionary science, especially Darwinism. He almost became a Christian, and his views were important in the development of Indian Christian theology. The Samaj eventually split into three sections, the Brahmo Samaj of India, led by Keshub, the Adi-Samaj, which was more Vedāntic, and the Sadharan Samaj, which stressed egalitarian reformism, and was more congenial to Vivekananda.

The Arya Samaj was founded in 1875 by Dayanand Sarasvati (1824–83). We characterize it as a reassertive response to secularization; its primary goal was to restore the Hindu tradition to an imagined pristine Vedic character. As a Vedāntist who believed in the infallibility of the Vedas, Dayanand was obliged to adopt a very arbitrary and selective approach to scripture. Thus, for example, Agni, taken as the name of God, means 'giver and illuminator of all things'. But as fire it means 'fire which give victory in battle', that is, guns. Dayanand believed that all modern scientific discoveries could be read back into the Vedas by this technique.[19]

Dayanand's selection of authoritative Vedas was arbitrary, but his basic argument that they are a revelation from God because of their consonance with nature, and therefore the fountainhead of all true science and religion, was more profound.

He firmly believed in reincarnation, which he interpreted deterministically to rule out both divine miracles and astrology – which he poked fun at, though he never managed to condemn it outright. Thus in an exposition of the laws of Manu he asserts:

> A youth should not marry a girl of... larger size... garrulous... or [with] inflamed eyes... nor one with the names of a constellation.... He should marry a girl with sleek proportionate limbs... with the gait of a swan or she elephant.[20]

The Arya Samaj had a much broader appeal than the Brahmo Samaj, and remains very active in social and educational fields. The reassertive anti-Western nationalism of today's Sangh Parivar can be traced to this brand of Hindu reformism. Dayanand's impassioned preaching made him many enemies and numerous attempts were made on his life.

The Ramakrishna Mission

Swami Vivekananda (1863–1902) founded the Ramakrishna Mission shortly after the death of his *guru*, Sri Ramakrishna (1836–86). Ramakrishna remained oblivious to Western science and education; in his mystical experiences he saw God primarily as the divine Mother, but also as Rama, Sita, Krishna, Muhammad and Jesus Christ – which convinced him that all religions are acceptable ways to God. Ramakrishna's disciple, Narendranath Datta, who is better known as Swami Vivekananda ('the bliss of discerning wisdom'), had received a Western education in Calcutta. Though from entirely different backgrounds, their fundamental message was the same: each person is potentially divine, and should work to release the latent inner power of divinity.

Vivekananda's published works are extensive; his *Complete Works* run to ten volumes. His fundamental philosophy was that of *advaita* Vedānta, whereby *Brahman* is the absolute and underlying ground of all appearance. For those with eyes to see, *Brahman* can be perceived as that which is real and unchanging within the manifold appearances which the senses encounter. In Vivekananda's words,

> Thinkers in ancient India gradually came to understand that... there was a unity which pervaded the whole universe – trees, shrubs, animals, [humans]... even God Himself; the Advaitin reaching the climax in this line of thought declared all to be but the manifestations of the One. In reality, the metaphysical and the physical universe are one, and the name of this One is Brahman.[21]

Science is therefore the study of the variations which have been manifested by *Brahman*, and since *Brahman* is ultimately one, all branches of knowledge must finally converge. Hence – following Vivekananda's line of thought – Jagadish Chandra Bose could deliberately target his research on the boundaries between hitherto different branches of biological science by investigating the possibility of pain in plants (he also had scientific reasons for doing this).

In terms of our three categories of response to secularization – the reassertion, adaptation and rejection of tradition – Vivekananda adapted *advaita* Vedānta with insights from Western science and philosophy. Some of these attempts were poorly formulated – he did not seem to be aware of the need to distinguish between cyclical Hindu cosmologies and the linear time scale of Western science, for example. He was familiar with Darwin's theory, which he compared to Sāṃkhya philosophy – one of the six orthodox Hindu schools (*darśana*s) of interpretation. According to this school, *prakṛti*, the eternal, unconscious potentiality of all being, rests in a state of equilibrium, composed of three 'strand substances', until *puruṣa*, the conscious intelligent self or essence, becomes present to it, whereupon the evolution of *prakṛti* from equilibrium occurs, creating a duality of subject and object. Vivekananda identified the Sāṃkhya term, *pariṇāma*, with the Darwinian process of natural selection.[22]

Vivekananda was not aware of Einstein's theories of relativity – he died in 1902 at the age of thirty-nine. But he incorporated the notion of space–time–causality into his Comprehensive Vedānta. Thus *māyā*, which, he argued, stands in the way of our apprehension of the One, and causes it to appear as Īśvara (the Lord), *jīva*s (selves) and *jagat* (the natural world), is made up of space, time and causality, which are interdependent. He did not say that space, time and causality are mind-dependent, but believed that the mind is conditioned by these factors, which are universal in the sense that they operate on all individual minds.[23]

At this point in his argument Vivekananda makes a passing reference to Kant:

> Those of you who are acquainted with Western philosophy will find something very similar in Kant. But I must warn you... that there is one idea most misleading. It was Shankara who first found out the idea of the identity of time, space and causation with *māyā*.[24]

But Kant and Shankara would have been asking different questions; the former, whether the 'thing-in-itself' can be known by the perceiving mind; the latter, whether sense impression conveys anything about reality.

Vivekananda's evolutionary ladder, whereby *advaita* Vedānta occupies the pinnacle of a triangle with Viśiṣṭādvaita and Dvaita lower down, must have been deeply offensive to non-*advaitin*s. His defence of reincarnation was imaginative, and his commitment to social equity continues to inspire young Hindus. The following two quotes are illustrative of his social concern:

> Why should I love everyone? Because they and I are one.... There is this oneness, this solidarity of the whole universe. From the lowest worm that crawls under our feet to the highest beings that ever lived – all have various bodies, but one soul.[25]

His view that an enlightened person should renounce salvation for the sake of others displays Buddhist influence: 'Do you think, so long as one *jīva* endures in bondage, you will have any liberation? So long as he is not liberated – it may take several lifetimes – you will have to be born to help him.'[26]

Like Roy, Vivekananda possessed a charismatic personality, and his appearance at the World's Parliament of Religions in Chicago in 1893 caused a sensation. Clad in the reddish robes of a *sādhu* (i.e. renunciate), he began: 'Sisters and brothers of America', whereupon the entire audience rose to their feet and cheered! 'I stand before you, representing a nation and a religion, in whose sacred language, the Sanskrit, the word "exclusion" is untranslatable.' (More cheers.) 'You have been told that you are sinners. You are not sinners, you are children of the living God. You are not sinners. It is a standing libel upon human nature to call a person thus.' (More cheers.)[27]

His audiences loved him, and his enduring legacy is to be seen in the number of small books of his teachings to be found in the rooms of students in major Indian universities. There are two statues in the grounds of the north campus of Delhi University, the Buddha and Swami Vivekananda.

Raja Yoga

Vivekananda's *advaita* Vedānta differed from that of Shankara in several respects. It was non-sectarian and inclusive with regard to gender and caste. It could be justified in terms of reason and science, and it could generate a practical, egalitarian ethic based on foundational Upanishadic texts (e.g. the *tat tvam asi* verse of the Chāndogya Upanishad). It also brought about a transformation of the understanding and practice of yoga.

Up to 1895 Vivekananda's main concern in the West was to explain and justify the Hindu tradition to his audiences. At about this time he began to concentrate his energies on building up a dependable and financially supportive following to whom he presented a more focussed form of universal teaching based on different types of yoga. His *Raja Yoga: Conquering the Internal Nature* was published in 1896.[28] In it he explains that 'as every science has its methods, so has every religion. Methods of attaining the end of our religion are called Yoga.'[29] There are four main types: karma yoga (the realization of divinity through duty and action), *bhakti* yoga (the realization of divinity through devotion to and love of a personal God), *raja* yoga, the attainment of divinity through control of the mind, and *jñāna* yoga (the realization of divinity through knowledge).

Vivekananda's adaptation of yoga – which is generally known as Modern Yoga – is largely based on the sections of the Yoga Sutras which are most practice-oriented. These are the *aṣṭāṅgayoga* (i.e. 'the yoga of the eight limbs'). According to Elizabeth de Michelis,

> [Vivekananda] brought to bear on his yoga elaborations very different systems of thought: from traditional Hindu lore to elements of Western science (mainly physics, psychology, anatomy and physiology); from modern philosophy (especially empiricism and idealism)...to the Neo-Vedāntic esotericism of the Brahmo Samaj.[30]

Vivekananda's reinterpretation of yoga for Western audiences subsequently branched into various forms of body–mind–spirit integrative training known as Modern Psychosomatic Yoga. Modern Postural Yoga subsequently emphasized postures (*āsana*) and breathing (*prāṇayāma*), whereas Modern Meditational Yoga stresses concentration and meditation.

Sarah Strauss summarizes Vivekananda's combination of *advaita* and yoga as follows:

> By emphasizing that traditional *advaita* Hindu goals could be achieved while continuing to participate in worldly life, using yoga as the vehicle, Vivekananda and his successors offered, quite literally, the best of all possible worlds to their followers.[31]

Vivekananda and reincarnation

Vivekananda's understanding of reincarnation in relation to science deserves to be considered in some detail because his arguments seem to be widely familiar in contemporary university circles. His works are readily and inexpensively available in publications by the Bharatiya Vidya Bhavan (e.g. *Bhavan's Journal* – the *Reader's Digest* of Hindu India). We shall consider later how recent scientific discoveries appear to pose problems for the Hindu understanding of rebirth; it is therefore important to mention the earliest attempts to adapt this crucial aspect of the Hindu tradition in response to the first encounters with Western science.

We have already noted Vivekananda's interest in Sāṃkhya philosophy. This provided him with a framework for understanding both biological and cosmological advances in science. Thus biological evolution was not so much a struggle for survival as a gradual development from one stage to the next:

> Our theory of evolution and of *ākāśa* and *prāṇa* is exactly what... modern philosophers have.... Belief in evolution is among our Yogis and in the Sāṃkhya philosophy. For instance, Patañjali speaks of one species being changed into another by the infilling of nature.[32]

He believed that humanity was latent in the ape, and the Absolute was gradually emerging from humanity. And since emergence is a natural progression from one species to a higher one, the Darwinian 'struggle for existence' is a misnomer:

> Our education and progression simply mean taking away the obstacles, and by its own nature the divinity will manifest itself. This does away with all the struggle for existence. The miserable experiences of life... are not necessary for evolution. Even if they did not exist, we should progress. It is in the very nature of things to manifest themselves.[33]

But this is not the same as the theory of natural selection according to which the struggle actually determines the character of the next member of the species.

Elsewhere Vivekananda couples this theory of heredity with reincarnation, though it is not clear whether or not he has fully understood the manner in which heredity is believed by biologists to operate:

> We have... gross bodies from our parents, as also our consciousness. Strict heredity says my body is a part of my parents' bodies, the material of my consciousness and egoism is a part of my parents'.... Our theory is heredity coupled with reincarnation. By the law of heredity, the reincarnating soul receives from parents the material out of which to manufacture a [human being].[34]

According to this way of looking at heredity, the role of parents is purely functional, and the real character and destiny of the individual is determined by the process of reincarnation. Unlike Dayanand, Vivekananda was able to find room for divine grace and a merciful God whose influence is consistently good. Thus although reincarnation is the primary determining factor in birth, God can mitigate its bad side without upsetting the total process:

> [God's] infinite mercy is open to every one, at all times, in all places, under all conditions, unfailing, unswerving. Upon us depends how we use it.... Blame neither [humanity], nor God, nor anyone in the world. When you find yourselves suffering, blame yourselves, and try to do better.[35]

Vivekananda was aware of the importance of reincarnation as a solution to the problem of undeserved suffering, and was able to find room for divine grace and personal devotion to a loving God. His arguments in support of reincarnation may be divided into three groups – those derived from Western philosophy, traditional Indian arguments, and arguments which appear to be of his own devising.

The main thrust of his concern to defend reincarnation was moral, and it seems likely that the climate which made it imperative for him to pose the problem of suffering in this particular way was provided by Christianity. He was also aware of the possibility of giving some sort of materialistic biological account of a situation which permitted people to be born with inherent deficiencies, but there is no suggestion that he saw science as a serious challenge to the doctrine of reincarnation. His severest criticisms were reserved for those who believed in a 'hideous, cruel and ever-angry God' – namely Christians and Muslims. He responded to the secularization of the Hindu tradition by adapting it with concepts many of which were borrowed from Western science. But he tended to overlook essential differences as, for example, between evolution according to the Sāṃkhya and the Darwinian process of natural selection. His conviction that true knowledge is to be found by searching for unity in diversity was expressed at a time when different branches of science in the West were moving closer together, and inspired Jagadish Chandra Bose and his scientific colleagues in their research.

Aurobindo Ghose

Unlike most of the personalities who have so far been discussed, Sri Aurobindo was never the leader of an organized reform group, but his influence was quite extensive. He can conveniently be classified together with the earlier Reformers who adapted Hinduism, and many of the arguments which he used in the process were derived from Western science.

Aurobindo Ghose (1872–1950) came from a sophisticated Bengali family who sent him to England for his studies. He rebelled against the excessive Westernization of his father, and took an active part in the beginning of the national struggle. In 1914 he founded a journal called the *Arya*, the aims of which were the systematic study of the higher problems of existence, and the formulation of a synthesis of knowledge harmonizing different religious traditions. Four years prior to this he had turned his back upon active politics and taken up a contemplative life at an ashram in Pondicherry.

The biographical details of Aurobindo's life are important because reaction against the West was an integral part of his outlook. He believed that in spite of all the technical mastery achieved by Europe during the nineteenth century, the direction which the West had taken was essentially wrong. How had the West gone so badly astray, and what sort of correctives were needed to redeem it? Aurobindo believed that the answer was to be found from an analysis of the way in which science had tried to progress. In order to conduct any experiment the scientist creates the idea of separateness from the object of enquiry. We must isolate a particular object or set of objects, and detach ourselves from them in such a way that any experiment can just as easily and successfully be conducted by someone else. But this procedure is essentially artificial because in reality no object is really isolated from other objects, and the scientist is an integral part of the surroundings.

Thus the understanding gained from scientific procedures is necessarily incomplete and the methodology of science can bring knowledge neither of the real world nor of God. And if this is true with respect to the physical sciences, how much more must it be true of biological enquiries into the nature of living organisms.

How then can reality be experienced and God known? Aurobindo claimed that the answer to this question came to him with burning conviction in the course of a prison sentence in Alipur gaol. While reading the Bhagavadgītā he became aware that what he had previously regarded as the religion of the Hindus was a universal message for all. The Hindus were a chosen people who had been entrusted with the truths whereby all might come to the knowledge of God.

Aurobindo's own writings are difficult to evaluate, and this brief summary is based upon G. H. Langley's biography *Sri Aurobindo*.[36] The impact of Western science upon Aurobindo's philosophy is most apparent in relation to his understanding of evolution. He accepted the principle of evolution, but argued that it is meaningless unless the goal is defined, and this is ultimate Reality. This Reality gives meaning to evolutionary processes, and it possesses a threefold character defined as *saccidānanda*:

> The conscious existence involved in the form comes, as it evolves, to know itself by intuition, by self-vision, by self-experience. It becomes itself in the world by knowing itself; it knows itself by becoming itself. Thus possessed of itself inwardly, it imparts also to its forms and modes the conscious delight of *saccidānanda*.... The Unknowable knowing itself as *saccidānanda* is the one supreme affirmation of Vedānta; it contains all the others or on it they depend.[37]

Aurobindo's concepts of involution and evolution are complicated and involve a special terminology. But his interpretation of *sat-cit-ānanda* (Sanskrit for 'being', 'consciousness' and 'bliss') is important, and will be referred to again when we discuss Brahmabandhab Upadhyay.

Rejecting the past

There do not seem to have been any significant all-out rejecters of traditional belief in the nineteenth century. The young Anglo-Indian Christian poet and teacher, Henry Derozio (1809–31), antagonized orthodox Hindus at the Hindu College in Calcutta with his rational views. He was dismissed in 1831 for reasons which included the charge that he did not believe in God. In protesting his dismissal he argued,

> far be it from me to say 'this is' and 'that is not', when after the most extensive acquaintance with the researches of science... we must confess with sorrow and disappointment that humility becomes the highest wisdom, for the highest wisdom assures [us] of [our] ignorance.[38]

As the century progressed a small number of people began to abandon their religious beliefs for a variety of reasons – science, rationalism, the influence of Bradlaugh and Ingersoll, for example. The Arya Samaj characterized such a person as a 'man without a *dharma*,... wading his way to liberalism through tumblers of beer'.[39]

There was no industrial revolution in India as in Britain, but new methods of travel and communication raised issues relating to religious practices. In 1844 when proposals were made for the building of railways, the East India Company objected because they felt that caste Hindus would not use them. But John Clark Marshman, a Baptist missionary, obtained a ruling from a body of orthodox Hindus that a pilgrim could travel by train without loss of merit, and that caste rules could be relaxed during the journey. In 1876, during a visit to India, Monier Williams (1819–1901), the Sanskrit professor at Oxford, was solemnly informed by a *paṇḍit* that ghosts were less in evidence since the advent of rail travel. On further enquiry it appeared that communication and travel had become so efficient that hardly anyone died without the customary burial rites by relatives, and hence there were fewer ghosts.

Rejecters of the Hindu tradition on account of science and technology became more apparent by the mid-twentieth century. They include Jawaharlal Nehru (1889–1964) and M. N. Roy (1887–1954). Of the various distinguished scientists whose achievements will be described in a later chapter, very few maintained that their work was not to some extent motivated by their religious and philosophical beliefs. Meghnad Saha was one such exception.

Science and the Muslims

The Delhi College played much the same sort of role among an élite group of Muslims in the 1830s as did many missionary institutions and the Hindu College (now Presidency College) among young Hindus, but it suffered a major setback in 1857–58 during the Sepoy uprising. During the latter part of the nineteenth and the early twentieth centuries Muslim learning was influenced by Syed Ahmad Khan (1817–98), who established what became Aligarh University, and Muhammad Iqbal (1875–1938).

Syed Ahmad Khan believed that interpretations of the Qur'ān should conform to the observed truths of science. He was wary of the ascription of 'miracles and supernatural attributes to an object or a person whom [Muslims]...consider to be holy or sacred. This is why [they]...have interpolated supernatural factors into Islam.'[40] Elsewhere he describes the limitations of Qur'ānic authority in relation to scientific knowledge as follows:

> The Qur'ān does not prove that the earth is stationary, nor does it prove that the earth is in motion. Similarly it can not be proved from the Qur'ān that the sun is in motion, nor can it be proved from it that the sun is stationary. The Holy Qur'ān was not concerned with these problems of astronomy; because the progress in human knowledge was to decide such matters itself. The Qur'ān had a much higher and a far nobler purpose in view.... The real purpose of a religion is to improve morality; by raising such questions that purpose would have been jeopardized. In spite of all this I am fully convinced that the Work of God and the Word of God can never be antagonistic to each other; we may, through the fault of our knowledge, sometimes make mistakes in understanding the meaning of the Word.[41]

If Syed Khan's adaptive approach to tradition encouraged a rational, critical attitude and a desire for knowledge in some Indian Muslims, Muhammad Iqbal gave them inspiration and an original philosophy – though there was a tendency to regard him primarily as a poet defending Islam. After teaching philosophy at the Lahore Government College, he went to Cambridge and Germany where he became acquainted with McTaggart, Bergson, Nietzsche and others whose influence is apparent in his poems.

Iqbal's basic assertion was that action is the key to existence, and that our destiny is to remake the universe: 'You are creation's gardener, flowers live only in your seeing.'[42] God desires the development of the self to its fullest capacity, but this can only happen through identification with the community of Islam, and through 'love', which is the *Sūfī* word for ecstatic devotion to God. Iqbal distinguished between the quietist strand of the *Sūfī* tradition, which he considered responsible for the decline of Islam, and the positive, optimistic side of *Sūfī* thought. His understanding of space and time is very striking:

> It seems as if the intellect of man is outgrowing its own most fundamental categories – time, space, and causality.... The theory of Einstein has brought a new vision of the universe and suggests new ways of looking at the problems common to both religion and philosophy.[43]

Einstein's insight that space and time form an inseparable continuum is hinted at on several occasions in Iqbal's poetry, as is McTaggart's view that temporal series are characteristics of appearances but not of reality, and Bergson's belief that everything is in a state of continual flux. In *The Secrets of the Self* Iqbal summarizes his concept of time as follows:

> The cause of time is not the revolution of the sun;
> Time is everlasting, but the sun does not last forever.
> Time is joy and sorrow, festival and fast;
> Time is the secret of moonlight and sunlight.
> Thou hast extended time, like space,
> And distinguished yesterday from tomorrow.
> Thou hast fled, like a scent, from thine own garden;...
> Our time which has neither beginning nor end,
> Blossoms from the flower bed of our mind.
> To know its root quickens the living with new life.
> Its being is more splendid than the dawn.
> Life is of time, and time is of life.[44]

Time is not the linear time which we say we feel and express in terms of past, present and future, nor is it the time of science, because the scientist must assign limits to both space and time. Real time is everlasting and infinite, and can be thought of as an attribute of God. Only in such time can the Self find its fulfilment.

Iqbal's concept of time is related to his activist attitude to life as a whole. If time is finite and limited, all that can be accomplished will ultimately disappear. Therefore life must look forward to a purposive irreversible goal:

> Every moment in the life of Reality is original, producing what is absolutely novel and unforeseeable. To exist in real Time is not to be bound by the fetters of serial time but to create it from moment to moment. Life is free creative movement in time.[45]

History is important because it is the arena of human activity, and the religious community has a vital role to play in shaping the future:

> The modern world stands in need of biological renewal. And religion, which in its higher manifestations is neither dogma, nor priesthood, nor ritual, can alone ethically prepare the modern man for the burden of the great responsibility which the advancement of modern science necessarily involves, and restore to him that attitude of faith which makes him capable of winning a personality here and retaining it hereafter. It is only by rising to a fresh vision of his origin and future, his whence and whither, that man will eventually triumph over a society motivated by an inhuman competition, and a civilization which has lost its spiritual unity by its inner conflict of religious and political values.[46]

The work from which this quotation has been taken, *Reconstruction of Religious Thought in Islam*, represents Iqbal's later outlook. Objecting increasingly to the *Sūfī* quietism which seemed to have brought Islam into fatalistic depths in the eighteenth century, Iqbal stressed the need for a type of biological renewal which would build a new élite into powerful units. However, this aspect of his thought belongs to the more recent past, and relates, in particular, to the genesis of Pakistan.

L. S. S. O'Malley claims that railways and steamers not only enhanced the popularity of pilgrimages among Hindus, but also increased the sense of solidarity among Muslims:

> In the case of Muslims railways and steamers have rendered ancillary service to religion by facilitating the pilgrimage to Mecca. Railways, steamers, the post and the telegraph have drawn them closer together and Islam has gained solidarity.[47]

A more detailed account of the changes that occurred in Islam in India during the nineteenth and twentieth centuries is given in Francis Robinson's recent publications, which are reviewed by Muhammad Qasim Zaman in the *Journal of the Royal Asiatic Society* (November 2004).[48] In this section we have attempted to highlight such changes as relate to science and science-based technology.

Rabindranath Tagore

Rabindranath Tagore (1861–1941) was the fourteenth of Debendranath Tagore's fifteen children. His father, though affectionate and scrupulously dutiful, was formidably aloof. His mother died when he was in his mid-teens, by which time he was a shy and introspective boy. He was seldom happy at school, and received no formal education beyond the age of fourteen. As a young man Tagore was not only unsocial, but gloried in being so. He was happiest when he lived on the family estates in central Bengal, usually on a houseboat on the river Padma. At the age of thirty-one, he described himself as follows:

> I am by nature a savage. Intimacy with men is absolutely intolerable to me. Unless I have plenty of room around me I cannot stretch my limbs, settle myself, and unpack my mind. I pray that [people]... may prosper, but they should not jostle me.[49]

Tagore's early years were steeped in brahmanical culture, with pilgrimages to the holiest of Hindu centres. By the age of twenty, his first volume of Bengali poems – written in isolation during his *zamindari* journeys – was hailed as a masterpiece by Bankim Chandra Chatterjee (1838–94), Bengal's fiery nationalist poet. He continued his *zamindar* role until 1897, moving permanently to Shantiniketan in 1901. Satyajit Ray, the Bengali film director, describes Shantiniketan as follows:

> It was a world of wide open spaces, vaulted over with a dustless sky, that on a clear night showed the constellations as no city sky could ever do. The same sky, on a clear day could summon up in moments an awesome invasion of billowing darkness that seemed to engulf the entire universe.[50]

Rabindranath's wife died in 1902, and a daughter the following year; subsequently his youngest son, another daughter and his only grandson died. His eldest daughter made a happy marriage, but also died. There is a beautiful passage about death in *Gitāñjali*, and a more philosophical reflection on the subject, written in 1911, and possibly influenced by Buddhism:

> unless we see death as conjoined with life, we cannot see truly. In our perception of everyday life we do not see death's role and we clutch at the world with all our might.... If we can only see death's true place then the burden of the world becomes lighter. But if we persist in seeing death as being quite separate from life then the pangs of bereavement will be severe. Death is the bearer of life, life flows forward in death's current: once this idea is properly grasped, our mind become free to see truth – within which there is no conflict. But as long as we persist in seeing life and death as fundamentally opposed, we feel attachment to the world. And attachment binds us, and makes us shed tears. All sins, fears and griefs spring from this.[51]

Tagore was a staunch Brahmo Samajist, and in 1911 assumed leadership of the major branch of the split which had occurred under Keshub Sen. At about the same time he became editor of the *Tattvabodhini Patrika*. His thinking began to move away from fierce nationalism to a more inclusive universalism. Thus the hero of *Gora* declares: 'It matters not whether we are good or bad, civilized or barbarous so long as we are but ourselves,' and 'In me there is no longer any opposition between Hindu, Muslim and Christian. To-day every caste in India is my caste, the food of all is my food.'[52] The key theme of *Gora* is the triumph of universalism over national egotism – a message which he applied to Europe on the verge of war: 'The problem of Europe is egocentric nationalism, a disease to be cured only by a universal ideal of humanity.'[53]

In 1912 Tagore published *Gitāñjali* ('offerings of song') in English, for which he was awarded the Nobel Prize for Literature the following year. Educated India went wild with excitement; on hearing the news, Tagore is alleged to have sighed and said, 'I shall never have any peace again.'[54]

Tagore founded a school at Shantiniketan, a rural retreat where his father often withdrew for meditation; in 1921 it became Vishva-Bhāratī University. The same year he returned from a European tour to find Gandhi leading a movement opposed to British influence in India, including the prevailing educational system. While acknowledging Gandhi's leadership, Tagore argued that Gandhi's rejection of the West was too undiscriminating and narrowly focussed:

> His call came to one narrow field alone. To one and all he simply says: Spin and weave, spin and weave. Is this the call: Let all seekers after truth come from all sides?... I am... anxious to reinstate reason on its throne. As I have said before, God himself has given the mind sovereignty in the material world.[55]

In his 1924 lectures in China Tagore argued that Asia needed above all to find her own soul in order that her universal message might be understood. His visit to China had a profound effect on him. It is also important to note his friendship with Brahmabandhab Upadhyay (1861–1907), who strongly influenced him. Upadhyay was a Roman Catholic who subscribed to *advaita* Vedānta, using it to bring together the Hindu tradition and Trinitarian Christianity.[56]

Tagore's travels took him to most parts of the world. He was extremely popular in Germany, less so in Britain and the USA. He objected to the way people in the USA identified Christianity with their way of life. In a letter to C. F. Andrews, he wrote:

> if Christ had been born again in this world he would have been forcibly turned back from New York had he come there from abroad if for no other reason, then certainly for the want of the necessary amount of dollars to be shown to the gate keeper. Or if he had been born in that country, the Ku Klux Klan would have beaten him to death or lynched him. For did he not give utterance to that political blasphemy, 'Blessed are the meek', thus insulting the Nordic right to rule the world, and to that economic heresy 'Blessed are the poor'? Would he not have been put in prison for twenty or more years for saying that it was as easy for the prosperous to reach the Kingdom of Heaven as for the camel to pass through the eye of a needle?[57]

In the same letter Tagore refutes the charge of pantheism, 'a word that has no synonym in our language, nor has the doctrine any place in our philosophy'.

With regard to science Tagore's views are complex and will be considered in more detail later. He subscribed to *advaita* Vedānta, but in *The Religion of Man* appears to move to a more Rāmānujist position:

> It is widely known in India that there are individuals who have the power to attain temporarily the state of *samādhi*, the complete merging of the self in the infinite, a state which is indescribable. While accepting their testimony as true, let us at the same time have faith in the testimony of others who have felt a profound love, which is the intense feeling of union, for a Being who comprehends in himself all things that are human in knowledge, will and action. And he is God, who is not merely a sum total of facts, but the goal that lies immensely beyond all that is comprised in the past and the present.[58]

In a conversation with Einstein in 1930 he argues that 'Science has proved that [a]...table as a solid object is an appearance, and therefore that which the human mind perceives as a table would not exist if that mind were naught.'[59] Elsewhere he states: 'If a particular man as an individual did not exist, [a]...table would exist all the same, but still as a thing that is related to the human mind.'[60] One might conclude that when he is arguing with Einstein as representative of Western science (as he sees it), Tagore counters what he perceives as Western dualism with a more Vedāntic insistence on unity.

Tagore's enduring legacy is evidenced by the remarkable range and number of acquaintances with whom he corresponded. Thus, for example, Susan Owen, the mother of Wilfred Owen, wrote to Rabindranath in 1920, describing her last conversation with her son before he left for the war which would take his life. Wilfred said goodbye with 'those wonderful words of yours – beginning at "When I go from hence, let this be my parting word".' When Wilfred's pocket notebook was returned to his mother she found 'these words written in his dear writing – with your name beneath'.[61]

Conclusion

The decision to introduce the English language as the medium of instruction for higher education throughout India had far-reaching consequences in virtually every aspect of public life. Not only was the need for sound science adduced – albeit misleadingly – as a justification for Macaulay's historic Minute in 1835, but Western science was seized upon by educated Hindus, Muslims and some Christians (e.g. Henry Derozio) as a means whereby tradition could be refurbished to meet the needs and challenges of the day.

Crucial to the Hindu response to Western education was the progressive reaction of the Bengali *bhadralok*, the leaders of which saw the need to focus their beliefs within the framework of *advaita* Vedānta, a standpoint from which they could counter Western religious and philosophical dualism with a more spiritually based monism. This was to prove particularly fortuitous with respect to the discoveries and the direction of science in the late nineteenth and early twentieth centuries.

Defining secularization as the process whereby areas of life and thought once considered to be under the tutelage of tradition cease to be such, we considered three logical types of response whereby tradition is *reasserted, adapted* or *rejected*, sometimes with the assistance of scientific or philosophical notions borrowed from the West. Ram Mohan Roy and the Brahmo Samaj were essentially agents of secularization (as was Yesudas Ramchandra of the Delhi Muslim College), whereas Dayanand Sarasvati and the Arya Samaj were reasserters. In fact the contemporary Sangh Parivar continues to reiterate the Arya Samajist assertion that all modern scientific discoveries are to be found in the Vedas.

Vivekananda and the Ramakrishna Mission, and Syed Ahmad Khan and the Aligarh Movement, responded to secularization by adapting their respective traditions with Western scientific concepts such as evolution and the need for rationality. Some of their detailed reasoning may have been flawed, but their imaginative arguments and self-confidence were an inspiration to those, like Rabindranath Tagore, who followed them.

Tagore's universal humanism carried forward the best features of Brahmoism – 'the pervasiveness of rational religion in all our life activities' – into a world increasingly fraught by national egotism.[62] He distanced himself from *advaita* Vedānta, largely to accommodate a more thoroughgoing devotional theism (and in the process moved closer to Rāmānuja's thought), but he none the less opposed the dualistic thinking which he perceived to be integral to Western science, while at the same time commending the researches of members of the Indian scientific community, such as Jagadish Chandra Bose, for their ability to bridge the gap between the living and the non-living.

3 Tradition redefined

It has been explained how *advaita* Vedānta came to occupy the central position in the Hindu tradition by the late nineteenth century, and subsequently. The main reasons were that it gave cohesion to the national movement (though it later also helped to create a gap between Muslims and the Hindu majority, leading to the creation of Pakistan); also it is a scholarly tradition based on reason and substantive texts; and it was seen by Christians as the high point of Hindu achievement. It also enabled a group of outstanding Hindu scientists to interpret their scientific work in terms of Vedānta.

But although the Hindu tradition became more focussed and less of a unity-in-diversity at the philosophical level, and reformed some of its less desirable practices, it retained much of its social *hubris*. This is illustrated with reference to a temple in north Delhi, after which there is a description of Vedānta. We have considered a modernizing movement within Indian Islam in the last chapter, and will not discuss it further because this would involve Bangladesh and Pakistan, which are beyond our scope. Christianity will be discussed with particular reference to Brahmabandhab Upadhyay, a close friend of Tagore, who developed an understanding of Christian theology based on *advaita* Vedānta. (The hundredth anniversary of his death is celebrated in 2007.)

A temple vignette

The following example illustrates how a modern temple in Delhi is able to cater for popular Hindu beliefs and practices while at the same time maintaining a traditional Vedāntist stance based on Shankara and his followers. Thus the temple *āchārya* is well-versed in Sanskrit and Shankara's philosophy and is able to trace his lineage through several generations. But this may not be apparent at first sight, and the casual visitor would need to spend some time at the temple and be familiar with Hindi to appreciate the combination of popular devotionalism and more orthodox tradition.

> Shri Giri Raj Kishore Sharma is *āchārya* of the Shri Madhav Ashram Hanuman temple (*mandir*) in north Delhi, located on the Kamla Nehru Ridge, three hundred metres from the main offices of Delhi University. The

ridge is a wild area populated mainly with trees, monkeys, a colony of peacocks, and a few courting couples each evening.

Ācārya Giri has lived at the temple with his family for thirty years. The temple was originally set up and dedicated to Hanumān by his *guru*, who was succeeded by a *sādhu*, who ran it for several years before leaving it to Giri.

Shri Giri possesses a doctorate in Sanskrit from Banares Hindu University and a master's degree from the Shankaracharya Ashram in Delhi. He is renowned for the quality of his chanting in Sanskrit, and much respected by the local academic community many of whom visit the temple daily. Kavita, his wife, is a qualified yoga and meditation (*samādhi*) teacher. She can teach *Haṭha*-yoga; once she went into a twelve-hour trance. On one occasion while meditating, she received a divine message to introduce Śiva into the temple. Śiva's *liṅgam* was therefore brought from Jaipur. The temple is now dedicated to Hanumān and Śiva, each occupying a separate building. On auspicious occasions the temple compound is packed with devotees; at Hanumān *jayānti*, for example, hundreds of local people, many of them children, are given a free meal.

Ācārya Giri and Kavita have four children, three daughters now in their twenties and late teens, and a son, Subodh, in his mid-twenties. Subodh recently married Reena, the daughter of a government officer in New Delhi – a well-educated and attractive young woman, appropriately two years younger and, of course, a brahmin. They have a daughter, who at the time of visiting (March 2006) is seven months old.

Subodh and Reena would like the two of them to live together in an urban apartment and hire a maid, but Subodh feels that he is needed at the temple – his mother is diabetic and his father spends a lot of time counselling visitors. Subodh does part-time work for the local branch of the BJP – the Hindu nationalist party which was defeated in the 2005 Elections. One of Subodh's sisters is married, the other two are still studying – one is reading politics, economics and astrology at a nearby college.

The festival of Holī is approaching and is marked by a full moon. I point this out to Subodh the previous night, and we discuss eclipses, which he believes to be caused by the demon Rāhu. I tell him that when I have a problem on my mind I look at the stars to get things into perspective. For his part he composes *ghazal*s – lyrical songs in Hindi (usually these are in Urdu). He possesses a letter from the editors of the *Guinness Book of Records* to say that they have published an entry stating him to have composed the record number of *ghazal*s. This was when he was younger. Back in the temple I am discussing Hindu philosophy with Shri Giri. He traces his line of tradition to the great Shankara himself. Among the points I raise with him is the interpretation of the phrase *sat-cit-ānanda*. He nods to Subodh, who explains that it means that 'you must be blissful under all conditions from your inner heart (*dil*).'

Before leaving the temple I meet Kavita returning through the evening shadows. She has been for a check-up following laser treatment at the All India Medical Institute for an eye injury. How is she? She replies that all is well – 'Thank God', she says (in Hindi), raising her eyes upwards.

The following comments are relevant. *Āchārya* Giri is at pains to stress his lineage, and that of his temple. He qualified at the highest level in Sanskrit at Banares Hindu University in the holy city of Varanasi. His teaching is in the tradition of the greatest of all Hindu teachers, Shankara, the chief exponent of *advaita* Vedānta.

The term *āchārya* is appropriate for Giri because he is an established and recognized teacher. Both the designations *āchārya* and *guru* are used by Vedāntins, but a *guru* is more of a spiritual teacher who may not be very learned. Women spiritual teachers are called *gurī*s. Giri is recognized as *āchārya* on account of his character as well as his knowledge. He is also a priest (*upadhye*) and a teacher (*paṇḍit*), but not a *sādhu*, because he has a family and is not a mendicant.

From the late nineteenth century onwards women increasingly assumed religious roles, though individuals such as the sixteenth-century *bhakti* saint–poetess Mīrābaī were already well known. Sri Ramakrishna achieved the first stage of *samādhi* under the tutelage of a woman tantric. It is therefore not surprising to find that *āchārya* Giri's wife Kavita plays such a major role in temple life. She is a qualified teacher (*gurī*) of yoga and a palmist, though not an astrologer. It was at her instigation that Śiva's *liṅgam* was brought to the temple and housed in a building next to Hanuman's shrine. Śiva would normally be associated with his son Ganesh, but Ganesh tends to be more popular in the south. In any case the gods are celestial manifestations of the One (i.e. *Brahman*), so there can be considerable flexibility as to how they associate with one another; some gods are more popular in certain regions than others (e.g. Kālī in Bengal).

Giri and Kavita's son Subodh is intelligent and well versed in temple lore. He is not a graduate, but speaks good English, partly because I helped him pass his class eleven examinations when I was lecturing at St Stephen's College from 1995 to 1998. His understanding of science is in some respects very limited (he believes in the eclipse demon Rāhu, for example), though when his mother suffered eye injuries which were aggravated by diabetes, he sought the best possible medical treatment at the All India Medical Institute in New Delhi.

Subodh's explanation of *sat-cit-ānanda*, based on his father's sermons, is very interesting. The threefold phrase is based on the Upanishads, and goes back to Shankara's followers, though not to the great teacher himself. The three words translate as 'being', 'consciousness' and 'bliss' or 'joy', and it is not surprising that Christian theologians have sought common ground between this threefold formula and the doctrine of the Trinity. (This will be considered in more detail in relation to Brahmabandhab Upadhyay.)

Subodh interprets the threefold formula in non-metaphysical and experiential terms: 'You must be blissful under all conditions from your inner heart.' This is sound advice, but with a bit of modification it could be stated so as to resonate even more with Christian Trinitarian theology, as follows: 'In the inner heart *ātman* and *Brahman* are bound each to each by intrinsic joy.' Finally, we note that the temple projects a convenient accommodation between Hindu populism, as represented, for example, by Hanumān and Śiva, and the austere monism of Shankara's *advaita* Vedānta, as expounded by *āchārya* Giri in his sermons.

Vedāntic origins

Vedāntic thought occupies the central position in the Hindu tradition. The term *vedānta* is derived from *veda* or spiritual knowledge, and *anta*, which defines both the end and the goal (like *finis* in Latin).

Vedānta is one of six major Hindu systems of thought, as follows:

1   Nyāya: logic or argumentation; based on the Nyāyasūtras, composed around the first or second century CE.
2   Vaiśeṣika: ontology or 'referring to the distinctions'; related to the Nyāya, but earlier, dated *c.* first century CE.
3   Yoga: often translated as 'discipline' (which can be mental as well as practical – it is important not to distinguish between theory and practice); dated *c.* first century CE.
4   Sāṃkhya: cosmology; although usually paired with Yoga, much of its cosmology and psychology was taken over by Shankara; dated second to fourth century CE.
5   Pūrva-mīmāṃsā: the earlier exegetical school dealing with the interpretation of ritual; dated *c.* first century BCE.
6   Uttara-mīmāṃsā: the later exegetical school based on the Upanishads and therefore called Vedānta; it is dated according to its particular exponent (e.g. Shankara *c.* seventh century CE).

These were the six main schools or *ṣaḍ-darśana*s, all of which claimed to be based on the Vedic corpus of scripture; of the six, the last two come closest to doing this. There were also lesser schools, such as the Śaiva-Siddhānta, the Kashmiri school and Tantra, which possess a vast literature most of which remains unread. But the *ṣaḍ-darśana*s have been studied in detail, and shape the outlook of most Hindus.

There were also the Materialists, known as the Lokāyatas or Cārvākas, whose founder is said to have been Bṛhaspati (*c.* sixth century BCE). They believed that all is impermanent and without distinction; thus immortality and caste have no ultimate significance and we must grasp life here and now. Though considered unorthodox, the Materialists were acknowledged as Hindus, which is why it is impossible to pin down any single belief of the Hindu tradition as pivotal. The tradition as a whole may be described as 'polycentric' and a 'unity-in-diversity', but it is not strictly correct to refer to it as 'Hinduism' in the same way that one refers to Buddhism or the Abrahamic traditions, because there is no corresponding monolithic entity.

The scriptural sources of Vedānta are the Upanishads, the Brahmasūtra and various texts such as the Bhagavadgītā. The Upanishads and their precursors were

36 *Tradition redefined*

composed over a lengthy period of time, and display major shifts of emphasis, which may be summarized as follows:

1   The Saṃhitās: these are hymns to the celestials (i.e. Gods (*deva*s) and goddesses (*devī*s)), which are manifestations of 'the One'. Well-known examples include Indra, Agni, Sūrya, Soma and Varuṇa (originally a 'high god' who faded into obscurity). The Saṃhitās, dated *c*.1500 BCE onwards, also contain details of scientific rituals and *mantra*s, powerful instruments which could be directed as occasion required. Thus in 1963 some *paṇḍit*s targeted *mantra*s at a meteor which approached dangerously close to the earth.
2   Brāhmaṇas: these are commentaries on the *mantra*s and are dated *c*.900 BCE. They internalize the sacrificial rituals of the Saṃhitās. The Śatapatha Brāhmaṇa, the 'Brāhmaṇa of a hundred paths', is the longest, and reveals much of the political and religious conditions of later Vedic times. Thus Agni is described as 'burning along this earth towards the east', an allusion to the Aryan practice of burning forests as they established settled agricultural communities from the Punjab into the Gangetic plains.[1]
3   Āraṇyakas: meaning 'that which pertains to the forest', dated *c*.800 BCE. These are adjuncts to the Brāhmaṇas, and their knowledge is for those who live in seclusion in the forests.
4   Upanishads: meaning 'to sit close by', because such proximity to the *guru* is essential in order to understand their deepest meaning. Their dates extend over a long period of time from *c*.800 BCE to as late as 200 CE. There may be as many as 200, though most commentators list about eighteen principal ones. The central teaching is that the self (*ātman*) is identical with the ultimate ground of Reality (*Brahman*), and that whoever realizes this attains release from suffering and rebirth.

It is important to appreciate the progression in understanding between the Saṃhitās (e.g. the Ṛgveda) and the Upanishads, reflecting the interiorization of the sacrificial ritual, in order to interpret the meaning of the texts. Many commentators such as Dayanand Sarasvati failed to realize the subtle changes in the meaning of words and concepts over periods of time. This is also true of sections of the contemporary Sangh Parivar.

Insofar that the Upanishads constitute the most substantive corpus of Hindu literature, we may affirm from them that the Vedic sacrifice lies within us, and the human heart is the altar. Foundational to the Upanishads is *Brahman* the great One, the undergirder of the whole universe. Who or what is *Brahman*? This is the crucial question at the core of the Upanishads. The Vedāntic tradition regards the Upanishads as revelation proper (or *śruti*, 'that which is heard'), but also attaches importance to the Brahmasūtra and the Bhagavadgītā, which belong to tradition (or *smṛti*, 'that which is remembered').

The Brahmasūtra contains about 550 aphorisms which set out the nature of *Brahman* and the means of attaining it. It begins with the Sanskrit text: *athāto brahma jyñāsā*. This translates literally as 'now – therefore – *Brahman* – inquiry', which may seem straightforward enough, though generations of commentators have written extensively on this single verse! Translation is sometimes made more difficult by Sanskrit rules of conflation (i.e. the *sandhi* rules). Thus *athāto* is a conflation of *atha* (now) and *ataḥ* (therefore). The disappearance of the short 'a' (which stands for 'not') in the process of joining words can complicate translation. A progression similar to that in the Upanishads occurs in the Brahmasūtra. Earlier sūtras had recommended ritual action as the means to fulfilment, whereas the Brahmasūtra stresses the need for knowledge (*jñāna* or *vidyā*) of *Brahman* as the means to ultimate liberation (*mokṣa*) from 'entanglement in the flow of individual existence' (*saṃsāra*).[2] The Brahmasūtra was composed around the beginning of the Common Era.

The Bhagavadgītā ('Song of the Lord') is the best known of the Vedāntic scriptures. It comprises a short interlude in the Mahābhārata in which Arjuna, a warrior preparing for battle against an army which includes his own relatives, explores his conscience in a discussion with Krishna, his charioteer, who ultimately reveals himself as Vishnu, the Lord of time. Krishna points Arjuna to the three liberating paths consisting of knowledge, selfless action and devotion to God. It was probably composed *c.*200 BCE.

The Upanishads, the Brahmasūtra and the Bhagavadgītā became the basis of the philosophy of Vedānta, for which Shankara, Rāmānuja and Madhva were the major commentators. Of these, we shall be especially concerned with Shankara's *advaita* Vedānta, which became the dominant philosophical framework of the nineteenth- and early twentieth-century Hindu Reformers.

The inverted tree

Shankara illustrated his beliefs with reference to a tree that is mentioned in the Katha Upanishad (section 2, 3, 1). The tree is an *aśvatta* or peepul tree and *Brahman* is its root. Since *Brahman* is superior to all, then nothing can be higher, so the tree must be upside down. The tree below contains many evils such as sickness, doubt and death. The leaves are the Vedas, which give true knowledge; the noise of the birds in the branches is the tumult of life, the world of differentiation, which is *saṃsāra*. The Bhagavadgītā also mentions this unusual tree.

Shankara (his name is variously written) was a South Indian brahmin from Kerala who lived in the second half of the eighth century. He died young, probably well before the age of forty. J. G. Suthren Hirst summarizes the other known facts of his life as follows: he left home at an early age to become an ascetic; he became a student of a famous teacher called Gauḍapāda and wrote commentaries on the Upanishads, the Gītā and the Brahmasūtra; he established religious centres at Śṛngerī (south), Puri (east), Dvārakā (west) and Badarikāśrama (north) – there are also claims that he established Kāñcī in the south; that he died in the Himalayas at Badarikāśrama.[3]

Shankara wore the ochre robes of a mendicant and travelled extensively, debating vigorously with opponents who held different philosophical views. These were mainly the Mahāyāna Buddhists, the Devotionalists and the Ritualists (or

Pūrva-mīmāṃsakas). Buddhism – anti-Vedic and anti-realist – was strong in the eighth century, though on the decline, dying out in most of India by the eleventh century. Devotionalism was always pervasive – our opening depiction of a temple in Delhi illustrates how Shankara's philosophy and popular belief can coexist, even though Shankara himself opposed the latter. The Ritualists remained strong until around 1600 CE, by which time the Vedāntist schools were dominant. They believed that the function of language is not to describe reality but to tell us what to do. Thus scripture is primarily about ritual, and questions about *Brahman* and the 'celestials' are secondary.

Shankara opposed all these schools of thought but, before discussing his views further, we shall mention another influential Vedāntin, Rāmānuja, who lived in the eleventh century – his dates are alleged to be 1017–1127 CE, which seems unlikely! Shankara and Rāmānuja remain the key figures in Hindu religious and philosophical thought; although Shankara's school was dominant among educated Hindus in the nineteenth century partly for political reasons, Rāmānuja's beliefs are as pertinent in relation to our preoccupation with science.

Rāmānuja was a Tamil brahmin who lived at a time when Shankara's philosophy was dominant. He was taught *advaita*, but disliked it and found another teacher who imparted a more devotional theology. Once his teacher was so pleased with his progress that he gave him a personal *mantra* to foster his spiritual development. Instead of keeping it to himself, Rāmānuja climbed the temple tower and shouted out the sacred *mantra* for all to hear. 'Why have you forfeited your salvation by doing that?' chided the teacher. 'Yes', replied Rāmānuja, 'I know I may go to hell for telling others what was for me alone. But even though I go to hell, they will be saved by the sacred knowledge.' The teacher was pleased with this response, and Rāmānuja eventually became *āchārya* of the Śrīrangam temple.

Both Shankara and Rāmānuja took the texts of the scriptures very literally. This is not to say that they did not recognize the metaphorical nature of some passages (e.g. a text which compares *Brahman* to an eagle). But their interpretations varied quite considerably. We note some of their general characteristics.

Shankara's inverted tree gives a graphic introduction to his thought. *Brahman* is the underlying ground of all appearance, the real and the unchanging lying within or beyond the appearances which human senses encounter. There is only one undifferentiated *Brahman*, and there therefore cannot be any duality between the self (i.e. the human subject) and *Brahman*, because *Brahman* truly is our self. Our aim in life must therefore be to realize that the self (*ātman*) is *Brahman*. A well-known passage in the Chāndogya Upanishad (section 6, 7, 1) equates *ātman* and *Brahman*: 'that self (i.e. *Brahman*) that *you* are' [emphasis mine].

It is important to note our use of the word 'appearance' to describe the manifestations of *Brahman* to the senses. The Sanskrit term *māyā* is often translated as 'illusion' which, whatever its precise dictionary meaning, is misleading. To emphasize this important point we quote Julius Lipner and John Bowker, as follows:

> With regard to our knowledge of the world, Śaṃkara is a realist.... The common supposition that Śaṃkara taught that the world is an illusion is a

much too superficial reading of his thought. For Śaṃkara, the world is as real as we are; only the fabric of worldly reality of which we are an integral part has no ultimate reality status.[4]

Māyā is often translated as 'illusion', but in fact *māyā* is the power of *Brahman* or of God to bring everything into the form in which it appears. Because everything is created and contingent, nothing is permanent... *Māyā*... is not the same as idealism in nineteenth-century Oxford, though both rely on the same underlying neurophysiology of perception.[5]

The classical example of *māyā* given by Shankara is the rope on a path onto which we superimpose our false belief that it is a snake.

Most Hindus (~90 per cent) are theists, and Rāmānuja was a theistic Vedāntin. Whereas Shankara concentrated his attention on the early, more monistic Upanishads, Rāmānuja cited later ones such as the Śvetāśvatara, which contain numerous references to Īśvara, the Lord, whom he called Vishnu Nārāyana. Rāmānuja's theology later became known as Viśiṣṭādvaita, 'a non-duality of particulars', or, more usually, qualified non-dualism. He paid particular attention to texts which refer to the *antaryāmī*, or 'inner controller'. Within everything, he argued, lies the inner controller, who is our *ātman* or self. Thus the whole world collectively and every being separately equals the 'body' of *Brahman*, the personal God. This is not the same as the Greek notion of a world-soul, or *anima mundi*. In acknowledging that we are the body of *Brahman* or *Brahma-śarīra*, Rāmānuja is giving more value to the material world than Shankara. The notion of *śarīra* means a relationship between God and the self implied by the pairs: controller and that which is controlled; supporter and supported; and principle and accessory. Does this mean that *Brahman* controls us? Not in the manner that *Brahman* as *antaryāmī* controls the sun, moon, and so on, but via the scriptures, the *āchāryas* and our consciences.

Shankara does not ignore scriptural verses in the earlier Upanishads which refer to a personal God, Īśvara. But Īśvara for him is sublated, transcendent and *nirguṇa* – the highest attributeless level of existence. For Rāmānuja there is always some differentiation between our individuality and God's. Neither thinker is pantheistic, and the term 'panentheism' utilized by Charles Birch, John Polkinghorne and others, is not really appropriate in this context.[6]

The differences between Shankara and Rāmānuja may be further illustrated with reference to their interpretation of the Chāndogya text: *tat tvam asi*. Shankara maintained that *tat* (that) is *Brahman*; *tvam* (our personhood) has a double meaning, (1) the individual ego or 'I', which is not *Brahman*, and (2) the underlying 'I' or *paramātman*, which is *Brahman*. Our task is to get rid of (1) and recognize and apprehend (2). Thus the external and cosmic *Brahman* and the inner *Brahman* (purified of self-centred individuation) will be one. For the same text Rāmānuja agrees with Shankara that *tat* is *Brahman*. But then *tvam* becomes the person as controlled, supported and accessory in relation to *Brahman*.[7] We are *Brahma-śarīra*, *Brahman*'s body.

40 Tradition redefined

The recognition or apprehension of *Brahman* is denoted by the term *pramāṇa*, which is the means for attaining true knowledge. Considering the six *darśanas* mentioned earlier (of which Vedānta is one), there are six *pramāṇas*. They are: direct sense perception (*pratyakṣa*), inductive inference (*anumāna*), verbal testimony (*śabda*), comparison (*upamāna*), presumption (*arthāpatti*) and non-cognition (*anupalabdhi*). The first of these, direct sense perception, is sometimes rendered by *anubhava*, and was translated as 'intuition' by the nineteenth-century Reformers and some Indian scientists to describe the process of scientific discovery (e.g. P. C. Roy).

The six *darśanas* mentioned earlier acknowledged varying numbers of *pramāṇas*. Shankara acknowledged the first three, but his attitude to them was complex, and we shall not discuss it here further.[8] Nor do we need to consider any of the other *darśanas*, except for the Sāṃkhya, the cosmology and psychology of which was taken over by Shankara, and utilized by some of the nineteenth-century Hindu Reformers on account of its superficial similarities with discoveries in the biological and physical sciences. The Sāṃkhya makes a fundamental distinction between selves (*puruṣa*) and matter (*prakṛti*). From a state of equilibrium of *prakṛti* three strand substances (*guṇas*) unfolded to produce the cosmos. Liberation is achieved by discriminatory knowledge (*sāṃkhya*), which is both theoretical and practical (i.e. yoga), which is why the Yoga *darśana* is associated with it.

The unfolding of the *guṇas* has been compared to evolution. This is not an accurate translation of the Sanskrit term *pariṇāma*, though it was used by several of the nineteenth-century Reformers. In any case 'evolution' was understood in several different ways in Europe. Similarly, the term for space (*ākāśa*) in the Vaiśeṣika *darśana* did not correspond to the notion of the ether wind, even before this was demonstrated not to exist.

It is unfortunate that the Hindu philosophical tradition, which has proved such a valuable resource historically in strengthening the self-determination of educated Hindus, has been more recently found wanting by Dalits. Dalits (i.e. 'the oppressed', the name they give themselves) are essentially the former untouchables at the bottom of the traditional caste system. They probably constitute a quarter of India's population, and in some states such as Uttar Pradesh their political clout is considerable. Most are Hindu, but they include Christians, Buddhists and members of other religious minorities who converted largely to avoid discrimination. Caste oppression has been and continues to be an unpleasant and unjust reality – we shall see later how Meghnad Saha rejected Hindu belief on account of his treatment by brahmins. But all religions at some time or another behave badly (e.g. the Spanish Inquisition, the religious wing of the American Right); this does not invalidate their essential teaching. In 1997, K. R. Narayanan was the first Dalit to become president of India.

Christianity vindicated

Following the end of British rule in India in 1947, a group of Indian Christians representing the major churches went to see Prime Minister Jawaharlal Nehru to ask what they should do under a predominantly Hindu government. That Nehru banged his fist on the desk may be apocryphal, but he certainly shouted at them: 'Vindicate yourselves!'

Twenty years later I heard Nehru's daughter, Indira Gandhi, take a similarly pragmatic, but more conciliatory, stance when opening the new wing of St Stephen's Hospital in Delhi. The Church of North India bishop began the proceedings with effusive gratitude to the Prime Minister for taking time to 'bless' the medical work of the churches. But Mrs Gandhi had no time for his speech. Taking the microphone from his hand she told the audience that Christians had no need of her to bless their social activities: 'You Christians have been running hospitals and schools better than we Hindus for many years', she snapped at the retreating bishop.[9] So have India's Christians – seen by many to be a product of the interaction between the Raj and India – vindicated themselves?

The 2001 census identified 24.1 million Christians in India (2.3 per cent of a population of just over a billion). But their presence is not by any means a consequence of the presence of the British. There were Christians in Kerala long before Christianity arrived in Britain; they are associated with the apostle Thomas, and are often known as Syrian Christians or Nestorians. The Mar Thoma church stems from this tradition.

Prior to and during the early stages of Indo-British contact, Christian churches were established by several European countries – the Portuguese on the west coast, the French in Pondicherry, the Danes in Bengal, for example. The Danish college at Serampore received a charter from the King of Denmark which allows them to award degrees in theology – neither the British Raj nor post-independence governments would permit theology to be taught in higher education.

Foreign incursions brought various branches of Christianity with them – Methodism, Lutheranism, Anglicanism, and so on. But it is to the credit of India's Christians that many of these have been able to bury those differences in order to set up the united Churches of South and North India (1947 and 1972 respectively). Problems remain, the most serious being pervasive attempts by fundamentalist Christian organizations based in the West to proselytize insensitively, using their enormous wealth to subvert the indigenous churches.

Christian schools, colleges and hospitals carry a weight of influence out of all proportion to the size of the Christian population, and their high standards in all subjects, including the sciences, mean that entrance to them is much sought after by Hindus and Muslims – so much so that Christians sometimes find it difficult to gain admission themselves. St Stephen's College in Delhi, Madras Christian College in Chennai, Edwardes College in Peshawar and several Roman Catholic institutions have consistently set excellent standards in higher education; top Christian schools include St Paul's, Darjeeling, Bishop Cotton school, Shimla and Tyndale Biscoe school in Srinagar.

India has yet to produce a Christian president. The nearest potential candidate was M. M. Thomas, the Mar Thoma theologian, who believed that religions find their dialogical focus in our common humanity. 'M. M.', with whom I collaborated in Bangalore in the 1970s, became Governor of Nagaland – an imaginative appointment considering that 98 per cent of the Nagas are Christian (mostly

Roman Catholic and Baptist). From state governor he might have become national president; sadly, he died. But although there are currently no Christians in line for such high office, there are many enormously talented ones in practically every national walk of life – Arundhati Roy, who recently won the Booker Prize, is a good example. (This is also true in Pakistan and Bangladesh.)

Few of the missionaries who dominated India's church life in the nineteenth and early twentieth centuries were interested in dialogue with Hindus and Muslims. Some Indian Christians exercised a degree of influence – Henry Derozio and Yesudas Ramchandra, for example. Keshub Chandra Sen was essentially a Unitarian with a deep respect for Christ, and under his leadership the Brahmo Samaj was very open to Christian influence.

As time went on some Christians with a commitment to inter-faith dialogue and engagement with intellectual issues such as new discoveries in science began to appear. The most important of these was Brahmabandhab Upadhyay, about whom we shall say more presently. Robin Boyd mentions P. Chenchiah (1886–1959), a South Indian lay theologian, who saw Christ as the fulfilment of the evolutionary process: 'He is *adi-purusha* of a new creation.... In Jesus, creation mounts a step higher... Jesus is the origin of the species of the Sons of God.'[10] Elsewhere he condemns the 'Latin isolationism' which lifts Jesus out of all human context, and maintains that sharp divisions in the boundary lines between God, humanity and nature will never take root in Indian soil.[11] His arguments are coloured with illustrations taken from biology and genetics.

Paul D. Devanandan (1901–62) was a leading Indian Christian who founded the Christian Institute for the Study of Religion and Society (CISRS) and played a major part in inter-faith and ecumenical activities based on the Ecumenical Institute, just outside Bangalore, and the United Theological College (next to CISRS in Bangalore). M. M. Thomas was also a director of CISRS, capably assisted by Richard Taylor, a distinguished sociologist and theologian from the USA. Paul Devanandan believed that before meaningful dialogue can take place between Christians and Hindus, it is necessary to identify the points at which secularization has rendered them most open to change. Robin Boyd summarizes Devanandan's four major areas of concern as follows:

> There are certain areas of thought in which it seems specially clear that traditional Hindu theology fails to provide the practical backing which is needed for life in a modern, developing country. These are the understanding of *personality*, where traditional conceptions of the *ātman* work against the ideas of freedom and development; the understanding of the meaning of *creation*, where Hinduism tends to underrate the importance of the material world and so fails to give adequate support to plans for raising the standard of living; the understanding of *history*, where Hinduism fails to provide a basis for the idea of purpose and planning; and *society*, where Hinduism... thinks mainly in terms of individual salvation and fails to envisage a 'transforming community'.[12]

This critique of the Hindu tradition within the context of modernity raises important questions in relation to science. Thus creation and history give meaning to the task of exploring the universe and the possibility of progress and technological development, and an understanding of the nature of human personality is necessary in order to describe our relationship to the natural world.

Where Chenchiah solved the problem of the relationship between God, humanity and the universe with the concept of evolution, Devanandan adopted a more empirical approach. In Sāṃkhya thought *puruṣa* is the creative principle which operates on *prakṛti* to produce the world. Devanandan believed that it is therefore ideally suited to convey the concept of human personality in a manner which does not draw unnecessarily sharp distinctions between God and the universe. He redefined *puruṣa* so as to describe the different facets of personal relationships – mutual encounter, purposive striving for the attainment of a common end, and the penetration of each personality by another.[13]

Unlike Vivekananda and Radhakrishnan, Devanandan did not adduce specific scientific theories in support of his redefinitions of traditional concepts. He was more aware of the climate of opinion created by science and technology than of any particular scientific advances, and he wanted to see both the Hindu tradition and Christianity develop in such a way that the relevance of both to modern India, and of each to one another, would be more clearly apparent.

Devanandan believed that, rightly interpreted, religion and science need not come into any kind of conflict, but that the crux of their relationship hinges, to a large extent, upon the nature of the attitude adopted by religion to creation. In practice Hindus accept the value of the material world and the importance of history as the arena for scientific progress, but from a doctrinal point of view he believed that *advaita* philosophy has not fully appreciated the situation:

> This new conception of history in the making in modern India will have to come to terms with the classical view about God and Reality that has so long held the field.... The Christian view of man as God's creature and of God as man's Creator, has provided the solution in Christian thought. But in Hindu thinking to accept the doctrine of the creation would be to do violence to the nature of God as the Absolute Being, who cannot be in any way involved in world life.[14]

Devanandan maintained that Radhakrishnan's description of the universe as being relatively real is not compatible with the idea of a personal God acting purposively in history. In contrast to the general evolutionary approach of the majority of Hindu and Indian Christian thinkers he believed that the future has already been realized in the present:

> Time is, as it were, shot through with eternity.... This is a world of *māyā*, a world which is both real and unreal, conditioned by time and shot through with eternity, the scene of human endeavour and the plane of Divine Activity.

But here the *sat-asat* nature of world-life is not understood in terms of Ultimate Reality but Final Purpose.[15]

The strength of Devanandan's arguments lies in the universality of his approach to non-Christian faiths and his interpretation of divine creation as implying dependence. He does not need to worry about the future consequences of evolution in terms of human progress since reality in its most profound form can be experienced here and now through faith in God. Hence there is no need for either rebirth or purgatory, and life beyond death can be conceived as a continuation of a life lived here and now according to the will of God. The soul may cease to exist at death, but the presence of God guarantees a type of eternal life which preserves some sort of personhood. For all people, whatever their religion, the unreal realm of *māyā* can be transformed into the real world of purpose and meaningful activity, as people align their own wills with that of God.

Devanandan and Chenchiah are representative of a group of Indian Christians who have been engaged in a profound and imaginative dialogue with Hindus about the implications of secularization for religious faith and valid reinterpretations of traditional forms of belief. Their work is not only immensely significant in the Indian context, but it also represents a challenge to theologians in the West. Their contemporaries and successors are few, and include such luminaries as Swami Abhishiktananda (Dom Henri le Saux), James Stuart of the Delhi Brotherhood of the Ascension, Bede Griffiths, and others. But we must give pride of place to the brilliant if highly controversial figure of Brahmabandhab Upadhyay.

The Hindu–Catholic fire-eater

In his recent determinative biography of Brahmabandhab Upadhyay, Julius Lipner heads each chapter with an allusion to fire.[16] Tagore described Upadhyay as 'a Roman Catholic ascetic, yet a Vedāntin – spirited, fearless, self-denying, learned and uncommonly influential'.[17]

Brahmabandhab Upadhyay (1861–1907) was born into a Kulin brahmin family in a village not far from Calcutta. After attending the local village school he embarked upon 'a rather unsettling westernized school career', in the course of which he developed an aptitude for both Sanskrit and Euclidean geometry.[18] The dominant religious influences on his home were Sakta goddess worship and Vaishnavism – with occasional visits from an uncle who had converted to Free Church Christianity. (Non-conformist Christianity was well represented in Bengal by Baptists in the tradition of William Carey (1761–1834), who, in addition to being a superb linguist, was also an expert botanist.)

Lipner divides Upadhyay's life into four phases as follows. First, there was the period we have just considered (1861–82), culminating in his first contact with Keshub Sen and the Brahmo Samaj. He had already met the young Vivekananda, two years his junior, at a Free Church missionary college – both participated in muscular sports – but it was Keshub in his role as leader of the reconstructed Brahmo Samaj who increasingly drew the young Brahmabandhab into his orbit and introduced him to his own mentor, the venerable Sri Ramakrishna.

Upadhyay was attracted by Ramakrishna's beliefs in the motherhood of God and the universality of religions, and by Keshub's attachment to the person of Christ and his attempts to reconcile Hindu and Christian beliefs. He also stopped wearing the sacred thread signifying his caste status, a practice he was to take up again later.

By now, the second phase of Upadhyay's life (1882–91), he was working as a Brahmo in Sind, in what is now Pakistan. Here he clashed with the Arya Samaj, which had become extremely strong in the northwest, with Lahore as its centre of operations. But he became increasingly fascinated by Christianity and started to move away from the Unitarianism of Keshub towards the Trinitarian view that Christ was both man and God. He was baptized by an Anglican clergyman in 1891.

Upadhyay's eventual choice of the Roman Catholic church was partly to do with the Catholic understanding of divine grace operating outside the perimeters defined by most Protestant missionaries. He was also attracted by the Roman liturgy, devotion to the saints (especially the Virgin Mary) and clerical celibacy. On the occasion of his reception into the Roman Catholic church he chose the name 'Theophilus' ('Theophilus' in Greek means 'friend of God', which translates into Sanskrit as 'Brahmabandhab'). He was then thirty years old.

The third phase of Upadhyay's life (1891–1903) covers the period from his conversion to Christianity to his re-evaluation of *advaita* Vedānta, during which time he considered himself to be a 'Hindu–Catholic'. Towards the end of this period (probably in 1903) he visited Oxford and Cambridge, where he lectured on *advaita* Vedānta. He was well received, especially in Cambridge where he made the acquaintance of the philosopher J. McTaggart. The final phase (1903–07) deals with his disappointment with the Roman Catholic church and his increasing involvement in anti-British politics. A talented journalist, his incisive attacks on British maladministration relating to Lord Curzon's partition of Bengal led to his arrest. He died from a painful form of tetanus before he could be sent to prison.

Upadhyay enjoyed science; he taught mathematics at St Patrick's school in Karachi. But it was in the area of religious philosophy that he made his most important contributions, thereby opening the way for a more positive understanding between science and religion. He set out rational arguments similar in some respects to those of Thomas Aquinas designed to establish the existence of God. His understanding of God – 'omniscient, omnipotent, spiritual First Cause, infinite in the sense that [God]...has "*no* parts" ', to which nothing can be added and from which nothing subtracted – is comparable with *advaita* Vedānta.[19] He quotes from Isaac Newton and G. J. Mivart, a biologist who converted to Roman Catholicism. 'In other words', Lipner concludes, he wanted to make his thesis 'in accord with modern scientific thinking...to strengthen its cogency'.[20]

Upadhyay's understanding of the difference between religious truths of reason and of revelation is illustrated in terms of the relationship between God's creative act and inner life, 'an inner life that revelation pronounces is trinitarian'.[21] But how far can the natural light of reason penetrate into the divine essence? This crucial question is answered in terms of an interpretation of the Vedāntic conception of the nature of God as *sat-cit-ānanda*, and of the production of the world as *māyā*.

46 *Tradition redefined*

Upadhyay's theological explorations were very much in line with attempts by the Roman Catholic church to justify natural theology in terms of an appeal to reconstructed Thomism. The nineteenth century had seen attempts by positivists and idealists to challenge Christianity (especially in continental Europe) by claiming that only mathematics and the natural sciences can yield truth – and this uniquely via logic or verifiable evidence of the senses. In the light of such critiques the claims of natural theology are invalid, and it makes no sense to conceive of God or immortality. Immanuel Kant and David Hume were often appealed to in support of this sort of empiricism, which the Roman Catholic church tried to counter with the kind of arguments adduced by the First Vatican Council (1869–70).

Upadhyay paved the way for a philosophically credible understanding of natural theology within a framework that was still essentially Thomistic, but which also contained *advaitic* elements. Lipner draws attention to some of these in a remarkable lecture by Upadhyay entitled 'A short treatise on the existence of God', which was eventually published under Roman Catholic auspices after his death.[22] For a theistic Vedāntin searching for common ground with Trinitarian Christianity it is strange that Upadhyay paid so little attention to Rāmānuja; he mentions him occasionally in his essays. But his nationalistic inclinations, backed by the opinions of Western orientalists, seem to have convinced him that Shankara was the best exponent of Hindu philosophical theology.

In recent years there have been a number of publications mainly by Roman Catholic authors, attempting to justify Christianity in the light of modern developments in science. The most important of these are published by the Association of Science, Society and Religion, an interdisciplinary venture of Jñāna-Deepa Vidyapeeth of Pune, of which Job Kozhamthadam is the president, and the Institute of Science and Religion at Aluva in Kerala – some are published jointly.[23] These publications are important contributions by Indian Christians, mostly Roman Catholics, to a mutual understanding between science and religion. But they lean heavily on Western authors such as Fritjof Capra, and have much less to say about the scholars who have featured in this chapter, especially Brahmabandhab Upadhyay. Perhaps this omission will be rectified in the years following the centenary of his death in 2007.

Conclusion

We began this chapter with an example of the manner in which popular Hindu beliefs and profound philosophy can coexist in a modern temple. Our illustration also indicated how a young Hindu can seek the most up-to-date scientific treatment for a parent while at the same time believing in the existence of an eclipse demon. (More persistent questioning might have led to a change of mind, though he had no problem with his sister studying astrology!)

The origins and general characteristics of Vedāntic thought were discussed, with particular reference to Shankara, the main proponent of *advaita*, which was considered to have a special significance by Indian scientists from the end of the nineteenth century onwards.[24] A brief outline of Rāmānuja's philosophy was added because it offers an arguably even more fertile framework within which to understand science. Reference was also made to Sāṃkhya cosmology and notions such as *ākāśa* (space), *pariṇāma* (roughly, evolution) and the *pramāṇa*s (means for attaining true knowledge, which include words corresponding to 'intuition').

Islam underwent a certain amount of modification in the hands of Syed Ahmad Khan and Muhammad Iqbal. We noted this in the last chapter, but consider modern developments in Pakistan and Bangladesh to be beyond our scope.

A brief account of Indian Christianity was given, noting the substantial contributions to national life made by Christian schools, colleges and hospitals. Though initially dominated in most parts of North India by missionaries, Indian churches eventually produced capable theologians such as Paul Devanandan and Brahmabandhab Upadhyay who addressed issues raised by science, and were able to do so within an ecumenical framework which offers great promise for a deeper mutual engagement between religion and science than has so far emerged in the West.

4 Worldviews in encounter

We shall consider the development of science in India prior to the arrival of the British, and the state of science in Europe during the nineteenth century when the encounter between Western thought and Indian society was most thoroughgoing – largely, though not exclusively, on account of the introduction of higher education in English. The purpose of the first part of what follows will be to demonstrate that, although Indian science lacked an organizational base and an experimental methodology (the 'Baconian philosophy'), it was selectively very advanced compared with what was going on elsewhere. Lord Macaulay could not have been more wrong in his contemptuous estimate of the state of science in India.

Our discussion of science in Europe may seem unnecessary on the grounds that it has been described extensively elsewhere. But from the perspective of late nineteenth-century educated India what was happening in Europe was interpreted in a distinctive manner, and we must delineate the European debate between science and religion accordingly. For one thing, whereas the British educators, administrators and missionaries expected their subjects to imbibe only what was best from English sources, in actual fact many of them also read widely in French and German.

We shall confine our account of the development of Indian science as far as possible – partly for brevity – to actual scientific discoveries. However, it is important to recognize some of the historical controversies that surround them. The Aryan Invasion theory, for example – as it has been dubbed by Rajaram and Frawley – has been discredited.[1] The Aryans, so-called ($ārya$ means 'noble'), were probably not a racial group, and although they may have migrated into India from the north, there is no evidence of conquest. According to Rajaram and Frawley, there was

> an organic development of a Sanskritic based culture in India from Vedic roots along the Sarasvati river, which then reorientated itself and made a later flowering along the Ganga [i.e. river Ganges] to the east, just as the archaeological record reveals the Harappan–Sarasvati culture coming to an end and a new but related culture arising on the Ganga to the east.[2]

The Harappān–Sarasvati culture is part of the Indus Valley civilization which began in the third millennium BCE. An important feature of Rajaram and Frawley's analysis – shared by many other Indian historians – is that it demonstrates continuity from the late Harappān period to that of Vedic times. Recent discoveries of black and red artefacts provide strong evidence for this. However, we cannot endorse Rajaram and Frawley's chronology where it is not supported by archaeological data.

The development of Indian science from antiquity to the arrival of the British may conveniently be divided as follows: a period ranging from antiquity to the first contact with the Greeks; a period during which links were made with Greek, Arabian and possibly Chinese science; the Classical Period; and the Islamic period.

Early Indian science

There is archaeological evidence to suggest that the Indus Valley civilization, based on Harappā and Mohenjo-daro, knew how to prepare implements of bronze and other alloys, and that they used some sort of plough to maintain crops. Their urban style of living suggests a fair degree of engineering ability. Bridget and Raymond Allchin have suggested the possibility of early contact with Egypt, but there is no evidence that technical skills were transmitted from one to the other.[3] By the mid-second millennium BCE pottery, carpentry and weaving were in evidence, and horses were in use, though initially only to pull chariots – it took four horses to pull a chariot. Stronger animals which could be ridden appeared centuries later. By the seventh century BCE the centre of activity had moved from the Punjab into the fertile regions of the Gangetic valley.

Vedic literature went through various revisions until c.800 BCE, by which time it had assumed a canonical block of four Saṃhitās ('collections'). The first of these includes the Ṛgveda, which contains references to solar and lunar movements, suggesting a primitive system of astrology or astronomy – the same term *jyotiṣa* (derived from the word for light) is used for both. Three and seven were sacred numbers, and it has been suggested in Ṛgveda 3.3.9 that the total number of gods is 3339 because this number can be obtained by adding 3003 + 303 + 33. When a round number such as a 100 or a 1000 is mentioned, it seems to mean 'a number of'. Thus the warrior god, Indra, is said to have destroyed a hundred fortresses. Fractions existed, and basic addition, subtraction and multiplication could be performed. But division was more difficult, and only Indra and Vishnu could divide a thousand cows by three; they obtained a quotient of 333, plus a remainder of one![4]

Sites for sacrificial rituals require geometry, and the Śulbasūtras, which are part of the Kalpasūtras, include methods of constructing suitable areas by stretching cords around stakes.[5] They also contain the theorem of Pythagoras. The Kalpasūtras are part of the Vedāṅga, which strictly lies outside the Veda. Calendars were developed to determine the timing of sacrificial rituals. It is interesting to note that during this early period it was essential for the fire priest (*atharvan*) to perform the sacrifice in the company of his wife – it was much later that women were considered too inferior to take part in religious rituals.

A more detailed account of Vedic mathematics based on the Śulbasūtras is given by Rajaram and Frawley.[6] Much of this draws on the work of A. Seidenberg, who distinguished two major streams of Vedic mathematical thought, the algebraic and the constructive, locating their origins in the Śulbasūtras:

> If it could be shown that each of these has a single source – and there are many rather familiar facts that suggest this is so – and if, moreover, in both cases the sources turn out to be the same, it would be plausible to claim we have found the unique origin of mathematics.[7]

The earliest classification of the Indian sciences is found in the Chāndogya Upanishad, which was probably composed before the fifth century BCE. Six sciences are described as 'limbs' of the Veda, as follows:

1 *Kalpa*, ritual
2 *Śikṣā*, phonetics
3 *Nirukta*, etymology
4 *Vyākarana*, grammar
5 *Chandas*, prosody
6 *Jyotiṣa*, astronomy (including astrology)

Mathematics developed from ritual, prosody and astronomy. Claims have been made that zero was indicated by *kṣudra*, which means trivial. Astronomy paid little attention to the planets, though there is mention of Rāhu, a demon responsible for eclipses.[8]

There are references to the abstract concept of number in various Vedic texts; the Śatapatha Brāhmana refers to periods of twelve months and 360 days and nights. The Chāndogya Upanishad also mentions *rāśi*, a term which came to refer to arithmetical operations. Acoustics developed in conjunction with the recitation of scripture; a Vedāṅga on metrics mentions the seven notes of the octave.

Vedic cosmology was complex. The earth was flat and circular, and separated from heaven by a middle air or atmosphere, in which birds, clouds and gods were situated. The sun, moon and stars were in heaven, and their movements controlled the destinies of humans. An all-pervasive order governed and regulated everything – it was known as *ṛta*, which was foundational to the later notion of *dharma*. The Purāṇas, composed much later (500–1500 CE), vastly complicated this simple picture by introducing a saline sea between the Indian sub-continent and its surrounds, plus various oceans composed of treacle, ghee, milk and curds. It was this Purāṇic facet of Indian imagery – not even canonical – which aroused the contempt of Lord Macaulay, and encouraged him to press for the speedy introduction of Western science.

The Śatapatha Brāhmana and the Atharvaveda contain texts which describe the bones of the human body, and prefigure the anatomical works of Caraka and Suśruta, whose medical treatises appeared around the beginning of the Common Era. Early Indian medical literature contained no borrowed Greek terms except a few of an astrological nature; thus *horā* in Suśruta's Compendium refers to an omen which precedes death. Early Buddhist texts contain evidence of the beginnings of a systematic science of medicine.[9] There was a taboo on contact with dead bodies. Chemistry appears to have been associated with medicine and metallurgy, rather than with alchemy.

Astrology remains popular in India, and we will therefore consider its basic elements. Astrology reflects – albeit inconsistently – the fundamental Indian presupposition of the unity of all existence. The significance of the Moon as the decisive influence within this matrix of being has been explained by Mircea Eliade as follows:

> By its mode of being, the moon 'binds' together a whole mass of realities and destinies. The rhythms of the moon weave together harmonies, symmetries, analogies, and participations which make up an endless 'fabric', a 'net' of invisible threads, which 'binds' together at once [humankind], rain, vegetation, fertility, health, animals, death, regeneration, after-life, and more.[10]

Calculations were based on the position of a fixed point on the eastern horizon or the location of the Moon in relation to the Zodiac at the time of a child's birth. The twelve signs of the Zodiac were related to areas of life, as follows: Aries, health; Taurus, status; Gemini, brothers and sisters; Cancer, parents; Leo, children and education; Virgo, enemies; Libra, spouse; Scorpio, life span; Sagittarius, virtue; Capricorn, activities and career; Aquarius, income; Pisces, expenditure.

Much depended on the position of the Moon and planets. Jupiter, Venus, the full and waxing Moon, and, usually, Mercury, were auspicious, whereas the Sun, Mars, Saturn, the waning Moon, Rāhu and Ketu (associated with eclipses and quite rare) were almost always inauspicious. If, for example, at the time of birth, Jupiter and Venus were in Taurus and Cancer, Saturn was in Pisces and the waxing moon was in Leo, the general horoscope would imply a well-placed family background, a good education, and several children who would live beyond their youth. However, the presence of Saturn in Pisces would suggest considerable expenditure, which might mean that all the children are girls requiring dowries at marriage. Professional astrologers can predict with great accuracy – a compendium recently purchased on a street corner in Bangalore predicts that if the Moon is associated at birth with Mars, the child may later become a wine merchant or a seller of counterfeit articles!

Palm reading is closely associated with astrology; both reflect the view that all parts of the universe are ultimately related in such a manner that it is natural to expect the behaviour of a distant planet to be associated with the patterns of lines on a human hand, and both to be related to a person's future. But this does not mean that the future is completely determined because different schemes of calculation permit some flexibility.

Astrology did not come into conflict with astronomy. But it was sometimes criticized for other reasons. Between the fourth and third centuries BCE, Kauṭilya, chief advisor to the Mauryan king, Candragupta, and author of the *Arthaśāstra*, expressed considerable scepticism at the king's decision to plan a military campaign according to astrological advice: 'Success eludes the fool who consults too much

52 Worldviews in encounter

the constellations. Purpose is the stimulating star for the success of the object in view. Of what avail are the stars?'[11] But Kauṭilya was also a pragmatist; on another occasion he advises that astrologers should predict certain victory for the king just before the beginning of battle (presumably to boost the morale of the troops).

The *Arthaśāstra* contains a great deal of important information about mining, metallurgy, engineering, chemistry, medicine, botany and astronomy – planets such as Jupiter and Venus are described in much more mathematical and non-mythological terms than previously. Belief in the existence of four fundamental elements originated in India prior to the sixth century BCE. These were earth, air, fire and water, to which several schools of thought added space (*ākāśa*), which in the late nineteenth century was compared to the ether wind, the non-existence of which paved the way for the theory of special relativity. Early atomism was developed considerably in the Vaiśeṣika school, and to a lesser extent in the Sāṃkhya, both of which came later. The Jains also subscribed to this theory.

External influences

During the period stretching from the fourth century BCE until the Classical Period (300–1000 CE), India encountered the science of Greece, and, possibly, China. Contact with Persian and Arabic thought came later.

Megasthenes, a Greek ambassador who spent several years in India at the end of the fourth century BCE, acknowledged considerable skill in practical science, but was critical of the level of basic scientific understanding: 'They are skilled in practice (εργοι) rather than in argument (λογοι) and they accept as true what is said in the myths (μυθος).'[12] Indian texts on astrology and astronomy of this period show signs of Greek influence, but there are no corresponding similarities in the medical documents. A. L. Basham maintains that medical science improved as a result of increased interest in physiology on account of yoga and through contact with Buddhist monks.[13] In return for food, monks sometimes offered medical advice which tended to be more rational and less influenced by magic than that of Hindu practitioners.

Severus Sebokht, a seventh-century Syrian astronomer who lived at Kenneshre on the Euphrates, credits Indian mathematicians with using a notational system of nine digits. He refers to the 'science of the Hindus',

> their subtle discoveries in the Science of Astronomy, discoveries that are more ingenious than those of the Greeks and the Babylonians; their computing that surpasses description. I wish only to say that this computation is done by means of nine signs.[14]

There is no mention of the tenth sign (i.e. zero), but the tone is much more complimentary than that of Megasthenes.

The zero, denoted by *śūnya*, came to have great significance in Indian philosophical thought, and was interpreted as a null point similar to the point of origin of a system of coordinates in coordinate geometry. Betty Heimann describes it as follows:

> The Zero-concept is not only a mathematical discovery, but was originally conceived as a symbol of *Brahman* and *Nirvāṇa*. Zero is not a single cipher, positive or negative (growth and decay) but the unifying point of indifference and the matrix of the All and the None. Zero produces all figures, but it is itself not limited to a certain value. It is *śūnya*, the primary or final reservoir of all single shapes and numbers.[15]

Śūnya is the etymological root of *śūnyata* or 'emptiness', which is foundational to the philosophy of Mahāyāna Buddhism. It is not surprising, therefore, that the origin of the notion is hotly disputed. Takao Hayashi, for example, claims that the oldest evidence for zero as a symbol is found in the works of Varāhamihira in the sixth century CE.[16] In our judgement it first appeared in India and was later taken up in the Arab world. Some Arabic literature refers to mathematics as 'the Indian art'. The diffusion of Indian science into Persian and Arabic literature probably occurred in the Classical Period (i.e. from 300 CE onwards).

In addition to contact with Greece and the Arabic world there seems to have been sporadic contact between India and China. According to Joseph Needham,

> It is...notable that the Sūrya Siddhānta describes the [armillary] sphere as being sunk in a casing representing the horizon, thus following traditional Chinese practice.[17]

The Siddhāntas were composed around the sixth century CE, and some of what they contained relating to science and technology was of Greek origin. Additional Chinese influence is a possibility, though when and how it occurred is not clear.

Within India we have noted some Buddhist influence. Jain sources also contributed to the early development of Indian algebra. They included methods of deriving square roots and estimates of the total population of the world which demonstrate a remarkable ability to manipulate large numbers. Our source of information for this is W. E. Clark, who states that according to the Jain texts the population of the world is the product of two raised to the power sixty-four multiplied by two raised to the power thirty-two.[18] Clark gets this calculation wrong, but the Jain text correctly (but unrealistically) works it out as two raised to the power ninety-six! These same manuscripts demonstrate familiarity with fractions, square roots, arithmetical and geometrical progressions, simple equations, simultaneous linear equations, quadratics, second degree indeterminate equations and summations of complex series.

Commenting on the period of the early development of Indian science as a whole, Betty Heimann observes,

> *Ṛta* is the functional balance of already existent single phenomena of which each in its proper place functions in its own law of activity, and all of them

collectively balance each other in mutually retarding or accelerating, limiting or expanding rhythm.... In all the various branches of Hindu knowledge this functional cooperation of living organisms is assumed.... [Humanity], animals and plants are for the Hindu interrelated and of essentially the same nature; all are part of the cosmic immanent life-force.[19]

The closeness of human beings to the plant and animal kingdoms, the basic unity and interdependence of all things, and the underlying oneness and order expressed by *ṛta* are fundamental to Indian thought and are reflected in the early development of Indian science in a variety of ways.

Classical Indian science

By the fifth century CE Indian astronomy and cosmology had developed well beyond the level of the Vedāṅgas and Purāṇas. Most of the information of this period is known from the work of Varahāmihira and the Āryabhaṭīya of Āryabhaṭa (b. 476 CE). The major features of the Āryabhaṭīya may be summarized as follows: (1) every phenomenon in astronomy repeats itself after a certain period of time; (2) all calculations must commence from the beginning of the *yuga* when the five known planets were in line with one another; (3) all heavenly bodies have equal linear motion; (4) the angular motions of heavenly bodies differ; and (5) the planets move in irregular paths because their motion is disturbed by the attraction of moving points in the heavens.[20]

Āryabhaṭa believed that the Earth revolves around the Sun and rotates on its axis, but his calculations were based on epicycles. A. L. Basham is wrong to say that these were the same as Greek epicycles, which were usually the same size, whereas those of Āryabhaṭa differed in size from quadrant to quadrant.[21] However, the notion of a *yuga* – the recurring cycle of humanity – is distinctively Indian.[22] Āryabhaṭa also gave a reasonably accurate value for π as 3.1416. The Āryabhaṭīya represents the emergence of astronomical science from much of its early mythology. By the time of its publication astronomers knew, for example, the true causes of eclipses, and no longer believed that they were caused by the demon Rāhu – an early example of secularization?

Mathematics went from strength to strength between the fifth and twelfth centuries, and major advances were made by Brahmagupta (b. 628 CE), Māhāvira (b. 850 CE) and Bhāskara (b. 1150 CE). Two major sources of information about the progress of mathematics during this period are the Brāhma-sphuṭa-siddhānta, and the Dhavalā, a commentary on a ninth-century Jain work.

Brahmagupta and Bhāskara identified the mathematical properties of infinity as follows: (1) any number divided by zero becomes infinity; (2) infinity divided by any number is infinity; and (3) infinity subtracted from infinity is infinity.[23] The Dhavalā improved upon Āryabhaṭa's calculation for π, and subsequent works gave values accurate to nine and ten decimal places. By contrast the Greeks were unable to improve upon Archimedes' value of 22/7 for several centuries. Negative numbers, powers and coefficients were in use, and Brahmagupta knew how to multiply and divide positive and negative integers. Śrīdhara (b. 750 CE) could solve quadratic equations. In the eleventh century Śrīpati anticipated Europe by six centuries by solving the equation:

$$Nx^2 + 1 = y^2$$

Trigonometry was in use in the Śūrya Siddhānta (fourth century CE), and Varahāmihira could solve various equations involving sine and cosine functions. In the ninth century Vāchaspati (b. 842 CE) defined motion as the change of position of a particle in space, measuring the exact location of one particle relative to another with a three-dimensional system of coordinates. This paved the way for solid and coordinate geometry and elementary calculus.[24]

By the end of the Classical Period astronomy and mathematics had progressed considerably, and Indian astronomers were in great demand in intellectual centres outside India. Physics, however, did not make as much progress, though the general properties of light and heat were known, and glass making and polishing had become a fine art. Bhoja (b. 1050 CE) was familiar with the magnetic properties of certain materials, and believed that ships made with iron joints were likely to be attracted to their doom by magnetic rocks in the sea.[25] Chemistry continued to be associated with medicine and metallurgy, and alchemy became increasingly important as a result of Persian influence. The quest for transmuting base metals into gold was pursued with great enthusiasm.

To sum up the previous three sections: the first indications of a rudimentary knowledge of certain sciences were the result of Vedic preoccupation with ritual and the need for a calendar. Geometry, astronomy, astrology, simple arithmetic and Ayurvedic medicine (which may have been pre-Vedic in origin) were known to the Vedic Indians, and, as the emphasis on ritual decreased, geometry became less important and algebra and theoretical mathematics came to prominence. Medicine appears to have reached its peak with the works of Caraka and Suśruta, and chemistry, which was originally associated with medicine, became largely subordinate to alchemy. Physics scarcely progressed at all, and primitive atomism, which lacked any kind of experimental basis, found its fullest expression in the Vaiśeṣika system. Acoustics developed along fairly independent lines of its own.

The influence of Greek science was most apparent with regard to astronomy and astrology, though even in this area Indian astronomers retained their belief in a recurring *yuga*, and used distinctive epicycles of their own. Contact with Arabia, Persia and China did not lead to any significant changes in Indian science, except possibly with regard to alchemy. The Classical Period witnessed a renaissance in Indian astronomy, mathematics and philosophy, some of which enshrined concepts which were later taken up by the nineteenth-century Reformers. But from the eleventh and twelfth centuries onwards a decline set in, and Indian science marked time throughout the entire period in which the scientific Renaissance in Europe was taking place. The northern part of the Indian sub-continent came increasingly under the control of the Turko-Afghans and later the Mughals.

Science under Islam

From approximately 1000 CE, when Muslim incursions into northwest India began in earnest, until the end of the Mughal empire, Islam exercised a steady influence on Indian society. But for many Indians the first awareness of a new religion came not from the invading armies of Maḥmud and his successors, but from the *Sūfī*s, who settled initially in Sind and the Punjab, and eventually moved into Gujarat, the Deccan and Bengal.

Sūfī mysticism and asceticism were generally appreciated by the Hindus, and their willingness to dissociate themselves from political and religious Muslim orthodoxy was an additional factor in their favour. Romila Thapar has summed up the role of the *Sūfī*s as follows:

> The *Sufi*s often reflected the non-conformist elements in society, and on occasion even the rationalist forces, since their mysticism was not in every case religious escapism. Some opted out of society in order to pursue knowledge based on empirical observation... Nizam-ud-din Aulia for instance, followed an inquiry on the laws of movement which displays a remarkable degree of empirical thought.[26]

Through the *Sūfī*s many Indians became acquainted for the first time with Arabian thought, and learnt of new scientific advances. But during the period immediately prior to contact with the Turks, Indian science seems to have stagnated, and although representatives of the two great cultures readily acknowledged each other's scientific achievements, there was little fruitful partnership. Al-Bīrunī (b. 973 CE) was impressed by the mathematics and astronomy of the Indians, but complained of their inability to distinguish between good and bad science, and their lack of scientific methodology:

> They are in a state of utter confusion, devoid of any logical order, and in the last instance always mixed up in silly notions of the crowd. I can only compare their mathematical and astronomical knowledge to a mixture of pearls and sour dates... or of costly crystals and common pebbles. Both kinds of things are equal in their eyes since they cannot raise themselves to the methods of a strictly scientific deduction.[27]

Al-Bīrunī was in India in 1030 CE, and he was able to experience Indian society as it existed shortly before Turkish dominance. Six hundred years later François Bernier, a perceptive Frenchman, offered his own verdict upon the state of Indian science under the Mughals:

> In regard to Astronomy, the *Gentiles* [i.e. Hindus] have their tables, according to which they foretell eclipses, not perhaps with the minute exactness of *European* astronomers, but still with great accuracy. They reason, however, in the same ridiculous way on the lunar as on the solar eclipse, believing that the obscuration is caused by a black, filthy and mischievous *Deuta*, named *Rach*.[28]

Presumably Rach was Rāhu and popular superstition had kept the old idea intact in spite of attempts by Bhāskara and others to explain the true cause of eclipses. Even so, it is strange that Bernier was so little impressed, and it must be concluded either that he was very prejudiced or that Indian science really had declined very seriously. Of anatomy and geography he observed: 'It is not surprising that the *Gentiles* understand nothing of anatomy. They never open the body either of man or beast.... In geography they are equally uninstructed. They believe that the world is flat and triangular.'[29]

But he was much more respectful of Ayurvedic medicine. He was also fascinated by the hold exercised by astrology:

> Here too is held a *bazaar*.... Hither, likewise, the astrologers resort, both *Mahometan* and *Gentile*. These wise doctors remain seated in the sun, on a dusty piece of carpet, handling some old mathematical instruments, and having open before them a large book which represents all the signs of the zodiac.... Silly women, wrapping themselves in a white cloth from head to foot, flock to the astrologers, whisper to them all the transactions of their lives, and disclose every secret with no more reserve than is practised by a scrupulous penitent in the presence of her confessor. The ignorant and infatuated people really believe that the stars have an influence which the astrologers can control.... The whole of *Asia* is degraded by the same superstition. Kings and nobles grant large salaries to these crafty diviners, and never engage in the most trifling transactions without consulting them.[30]

Both Muslims and Hindus practised astrology in order to match horoscopes prior to a wedding. Some still do. There are a number of Sanskrit translations of Arabic works of astrology.

Education was strongly encouraged by the Tughlaks, and the Muslim schools were financed by the State. But the emphasis was upon the learning of Muslim law, and relatively little attention was given to science. In some cases natural science was opposed as being contrary to the word of God. Zīa ad-dīn Barnī, an important political writer at the end of the thirteenth century, was highly sceptical of all forms of science which did not reflect the teaching of the Qur'ān:

> No other 'sciences' were allowed to be publicly taught...except Qur'anic commentary, the traditions of the Prophet, and law divested of all false interpretation – in short, apart from the 'sciences' which were based on the affirmation, 'God has said', and 'The Prophet has said', all other 'sciences' were banned.[31]

The Tughlaks saw no reason to dabble in the unknown, and their schools functioned largely in order to strengthen theological education. By the end of the Tughlak period there were up to a thousand *madrasa*s in Delhi alone.

At the end of the fourteenth century Mongol raids increased in frequency and intensity, and Tīmūr (Tamerlane) finally overthrew the Tughlaks, believing that God had designated him to punish them. Under the new Sultanate many more

58 *Worldviews in encounter*

Muslim scholars came to India, and some of the old ones fled from the main centres of learning to the provinces, carrying their knowledge with them. The emphasis of teaching became less theological, and more importance was given to the natural sciences. According to Romila Thapar, Indian medical systems became popular in western Asia:

> There was an interest in Indian and Arab learning, on both sides, in non-religious circles, and some intellectual exchange was inevitable. In medicine the interchange was particularly fruitful. Indian medical systems gained popularity in western Asia, and in reverse medical practices derived from these parts, called *yunani* medicine, were widely used in India, together with the earlier *Ayurvedic* system, as they are even to this day.[32]

Arabian techniques were not unlike traditional Ayurveda. There were four basic humours, and illness was believed to be caused by imbalance between them. S. H. Nasr maintains that the methods of Islamic medicine reflected a belief in a close relationship between humanity and the cosmos – a presupposition very similar to that of the ancient Indians.[33] According to A. L. Basham, the only change in Ayurvedic medicine which may be directly attributed to Arabian influence was the increased use of mercurial drugs, opium and sarsaparilla.[34]

Arabian science reached a peak of creativity around the ninth and tenth centuries, and from the tenth century onwards scientific works in Arabic began to carry the authority and prestige which previously had been accorded to Greek science. Jābir (b. 965 CE) and several other outstanding Arabian scientists made important advances in chemistry, astronomy, medicine and physics. Ibn al-Haitham used spherical and parabolic mirrors, and studied spherical aberration in a way which showed a careful experimental methodology, and he brought mathematics to bear upon a number of difficult optical problems. Al-Bīrunī carried out geodetic measurements, and determined latitudes and longitudes with some precision. In addition to introducing Arabian ideas into India he carried a treatise on Indian science, including the Indian system of numbers, into western Asia. During the eleventh century Omar Khayyam made contributions to algebra.

All these discoveries were made available to India, but until the end of the Tughlak period few major advances were made in the general level of scientific attainment. Under the Mughals, however, science was given official encouragement, and several emperors gave royal patronage to the study of natural science. Shah Jehan, for example, constructed huge astronomical instruments based on the theory of epicycles. Educational institutions were given more freedom with regard to what they taught, and the State encouraged them with gifts of rent-free land. S. M. Ikram points out that much of the educational activity was concentrated in the large urban centres where new ideas spread rapidly and met with a ready response.[35]

During the eighteenth century the educational curriculum was standardized, and although the natural sciences were represented, priority was given to philosophy and mathematics. Mughal political authority was by now on the wane, but the autonomy enjoyed by the *madrasa*s prevented a similar decline in educational standards. Although Mughal influence generally diminished, the further one went from the great northern capitals of Delhi and Agra, the existence of sea routes from the west coast of India to other Arabian countries stimulated the growth of intellectual centres in the Deccan. According to Ikram, astronomy, botany and geometry flourished in this area, and began to receive more attention than further north.[36]

Muslim science was based on presuppositions which had important points of affinity with the Indian view of the underlying unity of all things. According to S. H. Nasr:

> The Islamic classification of the sciences is based upon a hierarchy which has over the centuries formed the matrix and background of the Muslim educational system. The unity of the sciences has throughout been the first and most central intuition, in the light of which the difference sciences have been studied. Starting from this unarguable intuition of the unity of various disciplines, the sciences have come to be regarded as so many branches of a simple tree, which grows and sends forth leaves and fruit in conformity with the nature of the tree itself.[37]

By the time of Al-Bīrunī's visit in the eleventh century, Indian science had begun to stagnate, and the advent of the Turks did little to reverse this trend. The general level of education began to improve in the fourteenth century, and under the Mughals there was renewed interest in the natural sciences and medicine. But Indian science had passed its creative peak, and it was not until the second half of the nineteenth century that it began to gather momentum again.

European philosophy and science

Newton's pioneering work in physics took place against the background of a cordial and constructive relationship between science and religion – predominantly Christianity – in Europe. But the very success of Newtonian science fostered the emergence of a mechanical philosophy of nature with an accompanying mechanistic worldview. This led to reductionism (e.g. in biology), confusion over the notions of spirit and soul, and a general mathematization of the universe.

The mechanistic worldview encouraged scientism (i.e. the belief that the methodology of a particular branch of science can be applied in other areas), and, by reducing non-human living creatures to the status of machines, fostered cruelty towards them. Thus, according to Francis Bacon (1561–1626) – the first important British empiricist – the goal of science is to dominate and control nature, and knowledge of nature gives us power over it. René Descartes (1596–1650) believed that the purpose of science is to enable us to master and possess nature, though he avoided full-blooded materialism by postulating that our minds belong to a non-material realm. Though Immanuel Kant (1724–1804) came later, he was highly influential in intensifying these views and the prevalent Judaeo-Christian understanding that humans are destined to dominate the natural world; he believed nature to be a collection of irrational forces, and that holiness can only ever be achieved in society.

In spite of the devout theism of its originator, Newtonian mechanistic physics placed belief in God on a slippery slope, leading first to a divorce between science and religion, and then to alienation between them. However, by the second half of the nineteenth century, the mechanistic worldview began to be challenged in Europe from the direction of science. We need to be particularly attentive to this development, because it corresponded to a period of intense contact between Europe and North India, especially Bengal.

Mechanistic materialism was dominant in Germany during the nineteenth century, and provided fertile soil for the growth of logical positivism, a form of empiricism whereby theories are not only justified on the basis of facts obtained via observations, but are considered to have meaning only insofar as they can be so derived. The genesis of logical positivism in Vienna and Berlin may be attributed to a convergence of three philosophical streams, the empiricism of David Hume, John Stuart Mill and Ernst Mach, the methodology of science as developed by Henri Poincaré, Helmholtz, Boltzmann, and, later, Albert Einstein, and the symbolic logic and linguistic analysis of Ludwig Wittgenstein and Bertrand Russell.

Logical positivism banished all traces of metaphysics from philosophy and science, ignored culture, and challenged all the non-rational elements of religious experience. It therefore helped to maintain the gap between science and religion which had originally been fomented by the post-Newtonian mechanistic worldview even as the mechanistic outlook was itself being increasingly threatened by new scientific discoveries. Subsequent schools of philosophy – historicism and critical realism, for example – called in question the claims of positivism, and there were major controversies about the nature of scientific observations and their relationship to theory. The positivist view that religious statements which cannot be verified empirically cannot have meaning was shown to be untenable. Karl Popper eventually propounded the view that all science can ever offer is an approximation to the truth.

The eventual decline of mechanistic materialism and logical positivism has provided scope for more fruitful dialogues between science and religion, but for the time being we must turn our attention to the scientific and related discoveries of nineteenth-century Europe – especially the ones which were seen to have most significance from the viewpoint of educated, English-speaking India.

Darwin, evolution and progress

In this and the next section the impact of Darwinism and the work of evolutionary philosophers such as Spencer will be considered together with ideas with which they tended to be associated – the notion of prehistory and belief in progress, for example.

Another fundamental issue which stirred the imagination of nineteenth-century Europe was the new picture which began to emerge about the extent of the universe in space and time. As Darwinism raised basic questions for our understanding of our relationship to other kinds of life, and prehistory added an extra dimension to the human time-span, so discoveries in physics and cosmology provided a remarkable insight into the nature of the universe.

The best account of the scientific issues which stirred nineteenth-century Europe is Martin Rudwick's *Bursting the Limits of Time: The Reconstruction of Geohistory in the Age of Revolution*.[38] It is not proposed to summarize his arguments; we merely comment on the issues as they appeared to educated Indians towards the end of the century.

On the Origin of Species by Means of Natural Selection was published in 1859, but for more than half a century prior to its appearance the ground had been prepared from several directions. The discoveries by archaeologists of the earliest human tools were summed up in the *Bridgewater Treatises* which appeared in 1833, the same year that Charles Lyell published the third and last volume of his *Principles of Geology*. These contributions and the evidence of anthropologists extended the time-scale of human habitation of the earth far back beyond the beginning of Archbishop Ussher's chronology, and implied an even more remote age before it.

During the first half of the nineteenth century there were no significant reactions to the idea of prehistory, but public interest was aroused and held. *The Vestiges of the Natural History of Creation*, published in 1844, questioned the generally accepted chronology of the Bible, but failed to disturb the church to any great extent in spite of its immense popularity. Many theologically conservative scientists attacked it on scientific grounds and found weaknesses which were not difficult to exploit. But to the general public the idea began to get abroad that science was against religion, and it became essential for theologians to defend their doctrines in a manner which had not previously been necessary. However, neither the *Vestiges* nor occasional echoes of German biblical criticism produced any serious reconsideration until *Essays and Reviews* appeared in 1860.

Unlike the *Vestiges*, *Essays and Reviews* was not immediately intelligible to the general public, but it dealt a shattering blow to biblical conservatism among theologians and Church leaders. Between them these two publications seem to have achieved the same result at different levels – the former convinced the public that religion could not be evoked against scientific advances, and the latter produced the same conclusion in theological circles. Meanwhile archaeology, geology, and to a lesser extent anthropology, had opened up new and exciting possibilities for those who were willing to consider the scientific evidence without preconceived religious ideas. This was the background against which reactions to *The Origin of Species* took place.

Darwin's view that the animal kingdom has acquired its characteristics over a long time-scale was not original, and the theory of natural selection had been hinted at by continental biologists. He did not at first apply the concept of evolution to humans, and seems to have remained uncertain as to whether natural selection or acquired characteristics were the prime mover of change. But he had amassed a large quantity of evidence, and however hesitant his conclusions they offered an explanation of biological species which did not depend upon a mysterious 'vital urge' or act of creation.

Man's Place in Nature by T. H. Huxley and Darwin's *The Descent of Man*, which appeared in 1863 and 1871 respectively, applied the evolutionary theory to human beings, and Huxley succeeded in putting his ideas across with remarkable success to the general public. John Passmore has described Huxley, together with Tyndall and Clifford, as: 'The new scientific publicists... who, for the first time, took science to the working man, and with science, those heretical ideas about God and the soul which had previously been confined to a closed circle of "intellectuals".'[39] Huxley's success in this respect was an important factor in what eventually turned out to be a strongly emotional public reaction. Alvar Ellegård conducted a careful survey of the reactions of the periodical and public press to Darwinism, and his study makes it clear that many ordinary people felt deeply involved in the issues.

In 1861 Du Chaillu published an account of his travels in Africa and brought the existence of the gorilla to the notice of the general British public. According to Ellegård:

> Every newspaper and magazine carried stories of what was called [our] nearest relation, and though both Du Chaillu and most of the press expressly repudiated the descent theory, there is no doubt that precisely as Darwin had prepared the way for Du Chaillu's success, so Du Chaillu's book drew fresh attention to Darwin's doctrine, which was by now currently described as the 'ape theory'. Punch... had almost wholly ignored Darwin in 1860 [but] it made the gorilla and its relationship to [humans] one of the chief features of 1861.[40]

It is not easy to disentangle different strands in the debate, and much of what was said and written represents the beliefs and attitudes of professional scientists, church dignitaries and others. But for many people, the point at which Darwinism ultimately seemed to touch them most was that it presupposed a common origin for humans and animals. In this respect the public reaction was essentially one of psychological shock: 'Instead of Adam, our ancestry is traced to the most grotesque of creatures' – as the Monsignor observed in Disraeli's *Lothair*.[41]

The clash between Huxley and Bishop Wilberforce at Oxford in 1860 may well have pinpointed in a very clear way the exact nature of the conflict experienced by many ordinary people over Darwinism. Alec Vidler and Owen Chadwick both examined the different accounts of what was actually said on that occasion, and agree that Wilberforce's question to Huxley was essentially that, given his willingness to trace his descent through an ape to his grandfather, was he equally willing to make the same comparison on the side of his grandmother?[42] To ascribe human (i.e. 'man's') origins to the same source as an ape was bad enough, Wilberforce was arguing, but to do the same in the case of women was a direct assault upon one of the most cherished beliefs of Victorian society. Characteristic of several outbursts which echoed the same theme was Archbishop Manning's caricature of Darwinism as 'a brutal philosophy – to wit there is no God, and the Ape is our Adam'.[43] Vidler quotes another conservative who pleaded with the

Darwinists to 'leave my ancestors in Paradise, and I will allow you yours in the Zoological Gardens'.[44]

However, it would be incorrect to portray Darwinism only against the background of Victorian religious and social attitudes, and there were other notions in relation to which it needs to be evaluated. From the point of view of the nineteenth-century impact upon India the two most important ones are the so-called evolutionary philosophies, and the twin ideas – related but not identical – of progress and development.

Herbert Spencer was the best-known evolutionary philosopher, but his ideas were not consistent and are consequently difficult to summarize. In places, however, he appears to have made an important distinction between evolution from a lower to a higher form, and evolution conceived as a process whereby species change from a state of imbalance with their environment to one of harmony.[45] Spencer did not identify evolutionary change with development, and his emphasis upon the end-product of evolution as a state of equilibrium between the species and its environment would have readily agreed with the Indian notion that all things, by virtue of their underlying unity, are related. Behind and beyond the phenomenal world Spencer believed there is 'a Power which transcends knowledge', the study of which is the proper sphere of the religious enquirer or theologian.[46] Thus science was concerned with the phenomenal world, whereas philosophy probed relationships and degrees of generality between different sciences.

Like Darwin, Spencer had been stimulated in his evolutionary thinking by Lyell's *Principles of Geology*, and also by von Baer's researches in embryology. Neither *The Origin of Species* nor Spencer's extension of the evolutionary principle to virtually all branches of human thought answered a number of basic questions posed by natural selection, but biological and philosophical evolution was invoked to strengthen the belief of nineteenth-century Europe in progress.

The notion of progress had been particularly strong in eighteenth-century France, and the Abbé Saint-Pierre had argued convincingly for the application of human reason to social and political problems. John Baillie describes the eighteenth-century situation in France as follows:

> The passion for social reform, the determined application of [our] rational powers to this new field, confidence in the ability of law and government to bring about an immediate improvement in human affairs, and the combination of these proximate hopes with a still brighter vision of the more distant future – all these became naturalized in the French mind of this period, spreading also to England and Germany.[47]

But progress took as its starting point the present, and looked only to the future. On the other hand, development was a more flexible concept which could where necessary be more optimistic about the past than the future. In nineteenth-century England little distinction was made between progress and development, but whereas eighteenth-century France tended to opt for the former, Germany preferred the latter. (Scotland had close ties at that time with Germany.) French enthusiasm for progress spread to England at the end of the eighteenth century and found an early champion in Thomas Paine. But whereas the notion of progress in France had run counter to Christian belief, in England it took its place alongside it and so became one of a diffuse set of ideas which ultimately passed from England to India in the nineteenth century.

But these trends were apparent mainly in relation to the biological sciences. Meanwhile new and exciting developments were also taking place in physics and cosmology.

The nature of the universe

If the cumulative effect of discoveries in geology and anthropology was to greatly expand the nineteenth-century understanding of the extent of the universe in time, advances in astronomy did at least as much for estimates of its extent in space. In 1832 the astronomer Henderson proved the nearest star to be 24 billion miles away, and by the turn of the century astrophysicists had revolutionized pre-nineteenth-century notions of the nature of the universe and paved the way for Einstein's brilliant synthesis of what had once been regarded as separate scientific disciplines. In due course these discoveries were transmitted to India where, as will be seen, their significance was interpreted from a distinctive standpoint. Indian scientists and philosophers were particularly interested in two facets of nineteenth-century physics – the concept of ether, and the gradual unification of different branches of the physical sciences. We shall therefore consider European science from this perspective.

Newton had been the first person to give serious consideration to the possibility of the existence of an ethereal substance through which light could be transmitted across a vacuum. In his *Optics* he drew a parallel between a medium which enables heat to cross a vacuum and one through which light could be reflected and refracted. But his ether was static, an inert medium through which motions could be propagated. The weight of Newton's authority in favour of light corpuscles prevailed in England until 1801 when Thomas Young demonstrated interference with two slits and a coherent light source. Arago conducted polarization experiments with quartz crystals ten years later, and Fresnel and Young together developed the idea of light as a transverse wave motion moving in a direction perpendicular to the plane of vibration of the waves. Even before the time of Newton it had been known that light possesses a finite velocity, but it had not been determined with any degree of accuracy prior to Fizeau's measurements in 1849.

In order to explain polarization, Fresnel adapted Newton's ether and developed a mechanical conception of it as a universal fluid uniformly distributed in all space unoccupied by matter. The properties of Fresnel's ether were defined in terms of impressed motion and mechanical elastic deformation. However, he was unable to explain the behaviour of the earth and planets in relation to the ether through which they necessarily had to move, and but for Maxwell's successful attempt to unify optical and electromagnetic theory, the ether would probably have been quickly forgotten.

Meanwhile in electrostatics the evidence was pointing increasingly towards the continuing adequacy of Newton's ideas, and in particular of his notion of action at a distance. Coulomb had established an inverse square law of electrostatic attraction and repulsion similar to Newton's law of gravitation, and Faraday had measured the polarization of dielectrics. In 1819 Oersted showed that the magnetic properties of currents could be accounted for in terms of a law similar to Newton's and Coulomb's. Faraday then proceeded to draw mechanics and electromagnetism closer together with his equations governing the motion of magnetic poles round current-carrying wires. But at this point a question was raised as to how electrical forces managed to cross the space between objects, and in answering it Maxwell steered electromagnetic theory away from action at a distance to a new conception of the ether.

Maxwell published his electromagnetic theories shortly after the middle of the nineteenth century, and demonstrated mathematically that only one type of ether is necessary for both optical and electromagnetic phenomena. In place of Newton's action at a distance he postulated step by step propagation of waves through a medium capable of storing energy. Following Fresnel he made use of the notion of a transverse wave to describe the electrostatic and magnetic components of electrical phenomena, and showed that the longitudinal speed of propagation of such a wave in a vacuum is equal to the velocity of light. Thus Fresnel's ether was subsumed under Maxwell's more general and universal ether, and optics, electrostatics, magnetism and electricity were all accounted for by the same field equations.

But Maxwell said nothing about the connection between matter and ether, and was unable to explain optical emission and scattering, and the different velocities of light in material media and in a vacuum. But an even greater impasse was reached when Michelson and Morley failed to discover the presence of an ether 'wind' by comparing the behaviour of light waves perpendicular to and parallel with the earth's motion. The crucial nature of this measurement may be gathered from the fact that after Michelson's first attempt in 1881 the experiment was repeated in 1887, 1903, 1908, 1926, 1927 (three times), 1928 and 1930. It gave negative results in every case.

Meanwhile further progress had been made in electricity, and J. J. Thomson had carried out experiments to show that cathode rays could be explained as charged particles moving with very high velocities. In 1895 Lorentz had started to think in terms of negative electrons revolving around positive nuclei and emitting transverse electromagnetic waves, and this partially solved the problem of how electromagnetic waves originated.

But Michelson and Morley's crucial experiments had barred the way to any further development of the properties of Maxwell's ether, and it was only with the publication of Einstein's theories early in the twentieth century that real progress was achieved. Einstein dispensed with the ether, made the velocity of light a universal constant, and introduced the notion of a four-dimensional space–time continuum. Speculations about the ether and the problems raised by Michelson and Morley's failure to detect an ether drift were at their height during the last few years of the nineteenth century. Larmor published his *Aether and Matter* in 1900, and in 1908 Minkowski pointed out that space and time compensate each other and should therefore be considered as part of a four-dimensional continuum. From just after the turn of the century onwards the scientific world was dominated by Einstein and his theories of special and general relativity. We shall take up this theme again later.

Conclusion

Early Indian science was an integral part of a body of knowledge, in part religious, which progressed in some directions (e.g. mathematics) more than others. It influenced and was influenced by surrounding cultures less than has often been supposed, and it was selectively very advanced compared with what was going on elsewhere. But it lacked an organizational base and an experimental methodology. Contact with Europe introduced a new dynamic into educated India, and created a ferment out of which Indian science reconstituted itself along lines which were partly Western, but also shaped by indigenous cultural and philosophical factors.

Considering the nineteenth century as a whole from the point of view of the physical sciences, there was a general trend from the diversity exhibited by different branches of science at the beginning of the century towards unification at the end. Maxwell succeeded in combining electromagnetism and optical phenomena at about the middle of the century, and in so doing reinterpreted earlier concepts of the ether. Michelson and Morley's experiments in 1887 produced a temporary impasse which was only resolved with Einstein's total rejection of the ether, and the substitution of a four-dimensional continuum in which the velocity of light became a universal constant.

To the Indian scientists who began to distinguish themselves in experimental and theoretical research early in the twentieth century, Western science presented a curious picture. Some fastened upon aspects which appeared to have obvious parallels in ancient thought – the ether and evolution, for example. But on the whole such attempts were unconvincing, and more significant responses to Western science were those of P. C. Roy and Jagadish Chandra Bose, who saw in late nineteenth-century scientific advances an illustration of a fundamental Indian insight – the unity and interrelatedness of all things.

Not only were the actual achievements of the scientists an impressive demonstration of what Indian thought had always maintained, but the scientific philosophies of Spencer and Huxley were seen as preferable to Western Christianity. The following observation by a Bengali scientist published in the *Modern Review* in 1910, though coloured by nationalism, sums up the attitude of some educated Indians to what was happening in Europe:

> Let us see what effects the application of scientific method has produced on the most sacred of subjects, namely, our religious beliefs.... According to Spencer, behind all the natural phenomena there is the one Eternal Reality, to deny the existence of which makes the world utterly unintelligible but the

attributes of which are unknown and unknowable. This assertion, which horrified the theologians of Europe as heretical, appears to closely agree with the Hindu Spiritual Idea.... The great scientist-philosophers of the last century, Huxley, Tyndall and Spencer, were one and all impressed with the great mystery that underlies the phenomena of nature, and it is not too much to hope that if they were born in India they would have turned Vedantists with all their scientific knowledge. But unfortunately in England they were brought face to face with an inferior and dogmatic form of anthropomorphic religion with crude ideas of time and space and suitable only for the ignorant classes.[48]

Thus the same evolutionary ideas which in England had helped to drive a wedge between science and religion, when transferred to India, were seen to be consistent with the principles of Vedānta. But while this may have been true in general terms, detailed comparison between traditional Indian ideas and Western scientific concepts reveals important differences. *Ākāśa* (space) as understood in the *darśana*s, for example, may have had general points of affinity with nineteenth-century notions of the ether, but in view of the subtle differences between the ways ether was understood by Newton, Fresnel, Maxwell and Lorentz, it can hardly have been the same as all of them. But the nineteenth-century Reformers who tried to incorporate Western scientific ideas into their reinterpretations of traditional Hinduism seemed unaware of such fine distinctions.

The process of secularization whereby areas of life and thought, once seen as being under the tutelage of religion, came to be determined by secular and non-religious criteria was rapid and irreversible. Some examples have been tentatively given of secularization in relation to the early development of Indian science. But the introduction of Western education, philosophies and technology in nineteenth-century India produced a massive and thoroughgoing secularization far greater than anything that had preceded it.

5 Relativity and beyond

Referring to his scientific work, Einstein once observed, 'Things should be made as simple as possible, *but not any simpler.*' There are conceptual difficulties associated with our understanding of modern physics and cosmology, but these are not insuperable. One is psychological; doesn't it defy common sense that a rapidly moving clock should slow down? Another is linguistic; pairs of 'operators' which are 'canonically conjugate', for example. A third concerns the use of equations; but aren't these often little more than a convenient shorthand? (And lots of people who consider themselves ignorant of science learn to use shorthand very effectively.)

The two theories of relativity, the first dealing with uniform motion in a straight line (corresponding to Newton's first law), the second dealing with acceleration (corresponding to his second law), were published in 1905 and 1915 respectively. The earlier paper was only one of five produced within a few months while Einstein was working as a patents official at Bern in Switzerland; surprisingly, it was one of the others relating to the photoelectric effect which gained him the Nobel Prize in 1921.

The road to relativity

Prior to 1905, most scientists believed with Newton that all physical laws and the yardsticks and clocks used to measure them are invariant. If, for example, you are on a moving ship and you drop a tumbler on the floor of your cabin, it will fall and shatter in exactly the same way as if you were to do the same in your kitchen at home. In all normal situations Newtonian physics works well enough.

However, there are problems when we try to do such experiments with light, which by the end of the nineteenth century was known to be composed of two oscillating vectors, one electric, the other magnetic, at right angles to one another. (This was Maxwell's discovery – he expressed mathematically as four equations what we have just stated.)

Einstein's theory of special relativity endorsed Newton's view that the laws of physics take the same form in all inertial frames – for example the moving ship's cabin and the kitchen at home – but maintained that the behaviour of light is such that its speed remains exactly the same whether the source emitting it is moving or at rest. Of course if light is passing through, say, water, it will slow down, and if it passes through a prism or a rainstorm it will split into its various component colours (i.e. frequencies), but in a region devoid of matter (i.e. a vacuum) it will always travel at the same speed irrespective of the movement of the source emitting it. This defies common sense.

The focus of special relativity was therefore the *invariance* of both the laws of physics and the speed of light, and Einstein was quick to recognize that 'relativity' was a misnomer. It is therefore totally inappropriate for philosophers and theologians to elucidate theories of moral relativity (for example) with reference to Einstein's theories: 'the less they know about physics the more they philosophize', he once observed tartly.[1]

Einstein showed that the invariance of the laws of nature and of the speed of light are equivalent to the requirement that the coordinates of space and time used by different observers should be related by the Lorentz transformation equations. These contain the inverse function $\sqrt{(1 - v^2/c^2)}$ where v is the velocity of an entity and c is the speed of light. As v becomes very large and approaches the speed of light, the function becomes infinite, which is impossible, so we conclude that this cannot happen.

Newtonian physics assumes that the instruments we use to take measurements remain the same – that is common sense – but Einstein showed that under extreme conditions such as enormously high speeds this is not the case. The consequent transformation of time means that two events which are simultaneous according to one person will not be so from the point of view of another in uniform relative motion. Two observers in uniform relative motion will each see the other's clock run slowly – this is time dilation – and although the speeds necessary are too high to influence humans in normal situations, particles entering the earth's atmosphere from outside (e.g. certain cosmic rays) do show unmistakable evidence of this effect. But although time is dilated, sequences of events remain unaltered so that causation is unchanged. Also mass, momentum and energy are conserved.

The addition of time to the conventional space dimensions, length, breadth and height, requires a mathematical formulation, and this was given by Minkowski. According to his system the time axis is specified by ict, where t is the time, c is the speed of light and i is the square root of minus one ($\sqrt{-1}$). Points in this space are called events. Although the notion of a four-dimensional space–time continuum came into its own in general relativity – which deals with accelerations and therefore with gravity – we are still within the realm of Einstein's earlier 1905 theory.

We explain elsewhere how Newton's mechanistic worldview undermined the cordial relationship between science and religion and paved the way for positivistic and materialistic philosophies. Einstein's special relativity did not replace Newtonian physics, but rather subsumed it and Maxwell's electromagnetic theory within a more comprehensive framework. This weakened the philosophical basis for some of the undesirable side effects of mechanistic materialism such as reductionism, scientism, body/soul dualism, and what we describe as the mathematization of the universe.

However, Einstein's view that space and time are not independent invariables was revolutionary, and reverberated around the Western world, finding echoes

70 Relativity and beyond

among Hindus and Muslims in India. Einstein did not seem to appreciate the wide-ranging consequences of his theory. When in 1921 he visited England and the Archbishop of Canterbury asked him about the implications of relativity for religious belief, he replied that it was a purely scientific matter and had nothing to do with religion. (This occurred after the theory of general relativity had also been published.) But the distinguished German physicist, Erwin Schrödinger, did see a strong connection between special relativity and religious belief:

> It meant the dethronement of time as a rigid tyrant imposed on us from outside, a liberation from the unbreakable rule of 'before and after'. For indeed time is our most severe master by ostensibly restricting the existence of each of us to narrow limits – 70 or 80 years, as the Pentateuch has it. To be allowed to play about with such a master's programme believed unassailable until then, to play about with it albeit in a small way, seems to be a great relief, it seems to encourage the thought that the whole 'timetable' is probably not quite as serious as it appears at first sight. And this thought is a religious thought, nay I should call it *the* religious thought.[2]

Thus the more flexible understanding of time presupposed by special relativity does not devalue religious belief, and the Indian Muslim philosopher/poet Muhammad Iqbal was greatly inspired by Einstein's 'new vision of the universe'.[3] So was Tagore.

We have noted that special relativity does not influence the notion of causation. But other developments in physics soon began to appear to do so.

Quantum theory

Until 1900 it was believed that physical properties such as energy could assume any of a continuous range of values as prescribed by Newtonian physics. But this was not the case for 'black bodies' (i.e. objects which absorb all the radiation incident on them – stars, for example). Max Planck solved this problem by suggesting that radiation can only change its energy via discrete 'packets', which he called quanta.

Einstein applied Planck's insight to the phenomenon of photoelectricity, whereby certain metals give off electrons when light falls on them. (This is the basis of light detectors and television, and it was for this research with its enormous commercial implications that Einstein was given the Nobel Prize.) Einstein was bolder than Planck in that he decided that it was not just the exchange of energy between radiation and the recipient metal that was quantized, but light itself. Thus light possesses both wave and quantized corpuscular properties.

In the same year (1905) that Einstein published his work on special relativity and the photoelectric effect, he also published a paper on the properties of molecules in solution known as Brownian motion. He combined kinetic theory (whereby heat can be explained in terms of colliding billiard balls) with conventional hydrodynamics to derive an equation that showed that the jerky movements of small particles in solution vary as the square root of time. This was confirmed experimentally three years later, proving decisively that atoms really exist. Einstein produced two more groundbreaking papers the same year.[4]

In 1907 Einstein set out some novel views about the behaviour of solids at various temperatures, and two years later he formally proposed the idea of wave–particle duality. In 1911 Rutherford set out his solar-system model of the atom (a central nucleus with orbiting electrons), and two years later Niels Bohr 'blended classical and quantum physics imaginatively, even brilliantly,... but without fully satisfying anyone'.[5]

In essence Bohr showed that an electron circling a nucleus could only occupy an orbit for which its angular momentum was equal to $nh/2\pi$, where h is Planck's constant, and n is an integer (0, 1, 2, 3, and so on). Three years later, shortly following the publication of the theory of general relativity, Einstein demonstrated that a single light quantum (i.e. a photon) could stimulate an atom in such a manner that two quanta would emerge, resulting in light amplification. The laser, which stands for 'light amplification by stimulated emission of radiation', grew from this idea. However, noted Einstein warily, the enhanced emission 'leaves the time and direction of the elementary process to "chance" '. Was this a premonition of his later misgivings about the Uncertainty Principle? Possibly, but we hope we have said enough so far to show that Einstein (with Planck and Bohr) was the father of the quantum brainchild that he never fully came to terms with.

General relativity

Whereas the special theory of relativity deals with uniform motion (Newton's first law), the general theory covers motion between accelerated frames of reference (Newton's second law). Insofar as the most familiar type of acceleration that people encounter is the acceleration of free fall in the earth's gravitational field, general relativity brings us immediately up against the problem of gravitation. What is it?

Newton had interpreted gravity in terms of an attractive force between particles of matter with a force F given by:

$F = Gm_1m_2/x^2$

where m_1 and m_2 are the masses of the two particles a distance x apart, and G is the gravitational constant, which can be measured with great accuracy. Newton also showed that the external effect of a spherically symmetric body is the same as if the whole mass were concentrated at the centre. (This conception of a point mass proved adequate until the recent development of 'string theory'.)

General relativity explains such forces not in terms of attractions between masses but as a natural consequence of modifications in the geometry of space–time, causing it to become curved. It is this curvature of space–time, produced by the presence of matter, which controls the natural motions of bodies. In practice differences between general relativity and Newtonian gravitation only appear when the gravitational fields are enormously strong, as with black holes and neutron stars, or when extremely accurate measurements are made (e.g. the first experimental tests of this theory made by measuring the positions of stars near to the sun's disc during an eclipse).

In establishing general relativity Einstein introduced the principles of equivalence and covariance. The first of these deals with the issue of how, for example, the forces acting on a group of people in a car moving rapidly round a corner (centrifugal/Coriolis forces) relate to gravity. The relationship can be expressed using Riemannian space–time, which differs from the Minkowski space–time of the special theory of relativity in that the shortest distance between two events may not be a straight line but a curve called a geodesic. The extent of this curvature is given by the metric tensor for space–time, the components of which are solutions to Einstein's field equations.

Gravity can be understood in terms of space–time curvature because mass 'grips' space–time. According to Einstein,

> The general laws of nature are to be expressed by equations which hold good for all systems of coordinates, that is, [they] are covariant with respect to any substitution whatsoever (generally covariant).[6]

Thus the laws of nature are invariant despite the possible multiplicity of space–time representations of events.

Considering general relativity from a more philosophical standpoint, it can explain how differences in appearance or perception can be transcended. Clocks and measuring rods may be relative to frames of reference so that, for example, what appear as simultaneous events in one are not simultaneous in another, but there are deeper invariants such as the space–time interval between events. Representations of reality are determined by such invariants, which manifest themselves through all different reference frames.

After obtaining the field equations of general relativity, Einstein applied them to the universe as a whole, and in 1917 published 'Cosmological considerations on the General Theory of Relativity', the paper in which he added the cosmological constant to his field equations when he saw that the solution did not represent a static universe. Stephen Hawking summarizes this phase of Einstein's work as follows:

> Einstein's general theory of relativity transformed space and time from a passive background in which events take place to active participants in the dynamics of the universe. This led to a great problem that remains at the forefront of physics in the twenty-first century. The universe is full of matter, and matter warps spacetime in such a way that bodies fall together. Einstein found that his equations didn't have a solution that described a static universe, unchanging in time. Rather than give up such an everlasting universe, which he and most other people believed in, he fudged the equations by adding a term called the cosmological constant, which warped spacetime in the opposite sense, so that bodies move apart. The repulsive effect of the

cosmological constant could balance the attractive effect of the matter, thus allowing a static solution for the universe. This was one of the great missed opportunities of theoretical physics. If Einstein had stuck with his original equations, he could have predicted that the universe must be either expanding or contracting. As it was, the possibility of a time-dependent universe wasn't taken seriously until observations in the 1920s by the 100-inch telescope on Mount Wilson.[7]

But was Einstein really unwise to introduce a constant into his equations in order to render them compatible with a static universe? In the absence of evidence for an expanding universe until Hubble's experiments twelve years later, he can hardly be blamed for doing what he did. He later told George Gamow that the introduction of the cosmological term was the 'biggest blunder' he ever made in his whole life.[8] An unusual gloss on this incident is afforded by the correspondence Einstein occasionally undertook with children. An eleven-year-old girl had written to him: 'Dear Professor Einstein, I have been having problems with my maths homework at school.' Einstein wrote back: 'Yes, I do know how you feel; I've just been having problems with my mathematics.' He was referring to the cosmological constant!

Recent research on dark matter – intergalactic matter postulated to explain why galactic clusters remain gravitationally bound – suggests that something along the lines of Einstein's cosmological constant may yet need to be added to the relativistic field equations after all.

Bose–Einstein statistics

In 1924 Satyendra Nath Bose wrote to Einstein from India enclosing a paper entitled 'Planck's law and the light quantum hypothesis'. Einstein was sufficiently impressed by this to translate it into German and send it to a journal for publication.

Planck's law states that the energy of electromagnetic radiation is given by the product of the frequency of the radiation and an extremely small universal constant known as Planck's constant. Bose had derived this not from Newtonian electrodynamics, but by treating radiation as a gas made up of photons which can be studied statistically. But the application of statistics to small particles is based on assumptions about particle types, homogeneity, and so on, and Bose was the first to appreciate these distinctions.

To illustrate Bose's approach with an example – if three billiard balls need to be accommodated in two boxes, then the total number of arrangements is eight, that is, there are eight possible ways of combining the three balls in relation to the two boxes. However, if the three billiard balls are identical and therefore indistinguishable, then there are only four combinations. And if, in addition, there is a rule that not more than two balls can be accommodated at any time in the same box, then the number of possible permutations is only two.

The types of statistics corresponding to these three situations are Maxwell–Boltzmann, Bose–Einstein and Fermi–Dirac statistics, and the anti-overcrowding rule in the two boxes is the Pauli Principle. Heat molecules may be distinguished from one another and may crowd together, and therefore obey the first type of statistics. Photons, that is, light quanta, are indistinguishable and interchangeable but are not subject to the Pauli Principle, and hence obey Bose–Einstein statistics. And electrons not only cannot be distinguished from one another but are subject to crowding restrictions and hence obey Fermi–Dirac statistics. Particles which obey Bose–Einstein and Fermi–Dirac statistics are called bosons and fermions. These are classes of particles, and we shall meet them again when we discuss the Standard Model of atomic physics.

In terms of actual fundamental particles, by 1928 the only ones known were electrons, protons and photons (light quanta). Then in 1928 Paul Dirac proposed an equation for the wave functions of fermions which led to Chadwick's discovery of the neutron (which is a fermion) four years later. At about the same time Pauli postulated the existence of the neutrino to explain the continuous spectrum of electrons produced in radioactive decay (i.e. β rays – radioactivity had been discovered by Becquerel in 1896). These particles were conceived as fermions with half spin and zero rest mass.

From this time onwards fundamental particles proliferated to such an extent that, by the 1960s, we junior researchers working in laboratories felt that every time the top mathematical physicists couldn't solve an equation, they invented a new particle – and then told us to go and look for it. There seemed to be something conceptually unsatisfactory in particle theories and wave functions designed to simplify phenomena which were becoming more complex than the phenomena themselves. Noting that the neutrino was originally proposed to have zero mass, some of us at Manchester University jokingly observed that 'it doesn't exist, but it spins' (i.e. possesses an intrinsic angular momentum).

As time went on these proliferating particles assumed a more orderly existence, leading to the electroweak or Standard Model of particle physics, which came to prominence in the 1980s. Meanwhile back in the 1920s, Heisenberg and a group of outstanding physicists – but not Einstein – originated quantum mechanics and the Uncertainty Principle.

The Uncertainty Principle

In a letter written in 1926 to the German physicist Max Born, Einstein explained his attitude to quantum theory as follows:

> Quantum mechanics is certainly imposing. But an inner voice tells me that it is not yet the real thing. The theory says a lot, but does not bring us any closer to the secrets of the 'old one' (*der Alte*). I, at any rate, am convinced that *He* is not playing at dice.[9]

During the period of what is often called 'old' quantum theory, which ran essentially from Planck's law and Einstein's 1905 work on the photoelectric effect to the mid-1920s, when it was demonstrated that electrons on the surface of metals are scattered according to quantum rules (leading to the possibility of lasers),

Einstein was the dominant contributor. But he was uncomfortable with the notion of wave–particle dualism, and the statistical arbitrariness with which stimulated emissions emerged from atoms.

Bose–Einstein statistics mark the end of 'old' quantum theory and the beginning of new probabilistic approaches known as quantum mechanics. De Broglie had claimed in 1924 that all matter has a wave associated with it. Then in 1926 Erwin Schrödinger derived a wave equation which replaced the notion of orbital electrons having position and momentum with a wave function which predicted stationary waves of electron probability around the nucleus. Schrödinger's equation is a second-order differential equation the solutions to which can occupy physics undergraduates for up to two years!

In 1927 Werner Heisenberg proved that the position and momentum of an elementary particle can never be measured simultaneously with precision. In other words, the more you try to pin down the location of, say, an electron, in space, the greater will be the uncertainty in the product of its mass and velocity (i.e. its momentum), because the act of observing it (i.e. firing a light quantum at it) will disturb both its position and its momentum. Expressed mathematically, the product of uncertainty in position multiplied by the uncertainty in momentum will always be greater than a quantity based on Planck's constant.

What we have just stated is a particular case of a more general rule which would apply equally to, say, the energy and time of a particle. The more general case is derived from quantum mechanics – and here there is no way of avoiding either the mathematics or its characteristic jargon. Thus the Uncertainty Principle is valid for any pair of operators which are canonically conjugate with respect to the Hamiltonian (i.e. the observable associated with the energy of a system). Philosophers and theologians who do not understand the mathematical basis of this principle should refrain from drawing premature conclusions!

A consequence of the Uncertainty Principle is that it is impossible fully to predict the behaviour of a microscopic system. Thus the macroscopic principle of causality cannot apply at the atomic level. This put a proverbial cat among the atomic pigeons, and led to a rift between Einstein and some of his colleagues. But it was never insuperable, and it is important to recall the quote by Einstein with which we began this section: 'Quantum theory is certainly imposing. But...it is not *yet* the real thing' (italics mine).

Einstein was not alone in his unease about the Uncertainty Principle. Schrödinger was uncomfortable with it; in conversation with Niels Bohr he said: 'If we are going to stick with this damn quantum-jumping, then I regret that I ever had anything to do with quantum theory.'[10] Louis de Broglie was also unhappy with it, and tried several times to reconcile quantum mechanics with a more deterministic outlook. Einstein proposed several hypothetical arguments to discredit the Uncertainty Principle, each of which was countered by Niels Bohr. According to Bohr consideration must be given both to the system observed and to the measuring instruments, so that a change in the latter – such as the emission of the photon involved in the process of observation – even without any change in the system observed is considered to create a totally new situation.

In 1935 Einstein, together with Boris Podolsky and Nathan Rosen, proposed the Einstein–Podolsky–Rosen (EPR) paradox. This relates to the surprising consequence of quantum theory in that once two systems have interacted with each other, then a measurement on one system can produce an instantaneous change in the state of the other system, even if they are by then widely separated. Bohr had an answer to Einstein's challenge.

Another quantum conundrum concerned Schrödinger's cat. This poor hypothetical animal was put inside a closed box which also contained a radioactive atom with a 50–50 chance of decaying over a period of an hour, emitting a gamma ray in the process. If this happened, the gamma ray triggered the breaking of a vial of poison gas which killed the cat.

After one hour quantum theory tells us that the cat would be in an even-handed superposition of the states 'alive' and 'dead'. On opening the box the quantum wave packet collapses, and I find the cat either alive or dead. But such a state of affairs – whichever it is – can hardly have been brought about by my action in opening the box. Surely the cat would have known its fate long before I collapsed the wavepacket (still with its 50–50 superposition) by opening the box. The problem seems to arise because we are bringing together two situations, one about microscopic particles, the other dealing with macroscopic entities such as cats and people opening boxes.

These and similar thought experiments brought out some interesting though tentative philosophical speculations. Einstein was a philosophical realist, believing that the world has an existence independent of any observer; it is an entity in its own right and is populated by smaller entities such as protons and quarks. Bohr's position, according to which the role of the perceiving observer of atomic events was crucial, had much in common with philosophical idealism (i.e. the standpoint that assigns reality only to mental phenomena). But Heisenberg, a loyal member of Bohr's Copenhagen school, held a more flexible position. He once wrote:

> In the experiments about atomic events we have to do with things and facts, with phenomena which are just as real as any phenomena in daily life. But the atoms or elementary particles are not as real; they form a world of potentialities or possibilities rather than one of things or facts.[11]

So perhaps quantum objects do not carry classical quantities such as position and momentum (or energy or time), but they do carry the potentiality for such quantities. This may echo Aristotle's notion of *potentia* or Galileo's attempt to discard secondary qualities such as colour and taste in favour of primary qualities such as mass and shape. Perhaps *Brahman* can also be interpreted in terms of *potentia* for primary qualities.

This is speculative and anticipates later discussion, but it indicates the subtlety of the discourse between the participants in the debate about particle physics during the first half of the twentieth century. Einstein was a major and very creative participant in that ongoing debate, though he also had other things on his mind.

Towards a unified field

According to Steven Weinberg,

> One of Einstein's hopes for a unified theory was that it would provide an alternative non-quantum mechanical explanation of the atomic phenomena that had already been successfully accounted for by quantum mechanics.[12]

Apart from his periodic disputes with Niels Bohr and his associates, Einstein devoted the last decades of his life to the quest for a unified field theory which would bring together electromagnetism and gravitation in a manner that was consistent with general relativity. This was hardly unreasonable when one considers his groundbreaking achievement in formulating general relativity, and its remarkable experimental verifications.

In 1921 Einstein considered the idea that electromagnetism could best be understood as an aspect of gravitation in five dimensions. This enabled him to generate the equations of general relativity for the gravitational field and the Maxwell equations for the electromagnetic field, both in four dimensions. But the particles predicted by the extended theory turned out to be too heavy, so Einstein abandoned this approach. In more recent years 'string' theorists have utilized additional dimensions.

From the 1940s onwards Einstein considered the possibility that the metric tensor which determines position and time in a space–time continuum might not be symmetric. This generated an additional six fields, three of which he considered electric and three magnetic. The problem was how to relate these six new fields to the ten components of the symmetric part of the general tensor. He never worked this out.

Electromagnetism is now considered as part of a more general electroweak theory which unites it with the radioactive processes in which neutrons and protons turn into one another. These are short-range forces, whereas electromagnetism is long range like gravitation; nevertheless it is by incorporating the weak nuclear forces into electromagnetism that progress has been achieved. Another theory, known as the theory of quantum chromodynamics, accounts for the strong nuclear force that holds quarks together inside neutrons and protons, and neutrons and protons together inside nuclei. There are mathematical similarities between the electroweak theory and quantum chromodynamics, so it seems likely that these can be brought together before too long. When this happens, there will be a single unified theory which combines the electromagnetic, weak and strong nuclear forces. But gravitation will remain separate.

Indian scientists such as Satyendra Nath Bose were fascinated by Einstein's quest for a unified field theory because it seemed consistent with their Hindu belief that underlying the multiplicity of the phenomenal world, there is One, that is, *Brahman*. In 1951 Bose solved some of the wave equations governing the relationship between electromagnetism and gravitation, but this was only two years before Einstein's death, and the scientific world had lost interest in wave theories.

But one of Einstein's later achievements, stimulated by Bose's 1924 research paper, was the prediction that a gas of free bosonic atoms (i.e. particles with integral spin) could condense into a superfluid. Superfluids move without resistance, and can 'climb' out of a container in apparent defiance of gravity. Bose and Einstein studied the properties of helium, which change dramatically at a critical temperature just above absolute zero (known as the lamda [λ] point). They contended that as atoms in a domain close to absolute zero ($-273°$) approach standstill, their wave functions merge to form an entirely new form of matter known as the Bose–Einstein condensate. This was discovered in 1995.

The particle named after Bose, the boson, is a key component of the Standard Model of fundamental particles. According to this scheme there are two elementary classes of particle associated with weak nuclear interactions. These are known as W and Z bosons, and were first observed experimentally in 1983. The Higgs boson, first proposed by Peter Higgs, a British physicist, has yet to be observed experimentally. Higgs maintains that all space is permeated by a Higgs field comparable to an electromagnetic field with the property that all particles, as they pass through it, acquire mass. The principle of wave–particle dualism then requires a particle, the Higgs boson, to be the agent of interaction. The discovery of this elusive particle, if it exists, will probably be made by the Fermilab Tevatron collider in Chicago, or the Large Hadron collider in Geneva. The latter is due for completion in 2008.

A road less travelled

My own research in physics at Manchester University in the 1960s related to the strong (i.e. nuclear) and electromagnetic forces. We were using an electrostatic van der Graaf generator to induce nuclear reactions in specific elements which, in my case, included germanium, molybdenum and an isotope of carbon. Six million volts on top of a large sphere above our heads directed particles downwards on to small targets with which they would react to give off new particles whose energy we could measure – very precisely – by deflecting them horizontally onto emulsion plates.

The research was a fairly routine filling in of gaps in knowledge about the nuclear properties of chemical elements. It was exciting that we were using a research report given to me by Åage Bohr, son of the great Niels Bohr, ahead of publication. There was a lot of mathematics, and we used the newly installed ATLAS computer to perform distorted wave Born approximations – named after Einstein's friend Max Born. It didn't worry us that we were using different 'models' for experiments conducted in different parts of the Periodic Table – the optical model for carbon thirteen elastic scattering and Åage Bohr's more empirical model for middle-range elements such as germanium; had we been working with heavy elements such as uranium, we would have used the so-called liquid drop model. Were these three different models of the nucleus compatible with one another? We never gave it a thought.

At this time physicists were extremely self-confident, and saw no reason to listen to anyone who didn't work in the natural sciences. Religions were out of court altogether – in fact we sometimes speculated about whether or not our deadly six million volts could bring down the huge tower of the Church of the Holy Name, a mere fifty metres away on the Oxford Road!

Our mood of self-confidence was endorsed by the views of Fred Hoyle, the ebullient Yorkshire cosmologist, who, together with Bondi and Gold, had proposed the 'steady state' theory of the universe's origin. Roughly this was that matter is continually appearing in our universe to create the illusion of expansion; the rival 'Big Bang' school maintained that it had originated from some sort of primitive condition a long time ago. Hoyle rather enjoyed provoking the Christian establishment by arguing that his theory contradicted the Judaeo-Christian accounts of creation in the early chapters of the book of Genesis. It was all very entertaining; but then in 1965 two astronomers detected microwave radiation which could not be accounted for by Hoyle's theory. Hoyle subsequently collaborated with the brilliant Indian astrophysicist Jayant Narlikar.

Prior to working in particle physics, my fascination for astronomy had led to two internships, one at the Royal Observatory at Herstmonceux in Sussex, where I photographed a ninth magnitude minor planet called Interamnia, the other at the Jodrell Bank Radio Telescope (which was part of the department of physics at Manchester University), where I took part in an experiment to bounce radio waves off the planet Venus. This was a time when astronomy was moving forward in leaps and bounds: quasars were discovered in 1963 (these are extra-galactic sources of huge amounts of infrared and x-ray radiation – a few are radio sources); then came pulsars (which rotate and may be composed of neutrons); black holes (caused by the gravitational collapse of massive objects) came later and moved progressively into the centre of cosmological debate. It is a supreme irony that although Einstein totally repudiated the possibility of black holes, Freeman Dyson currently describes them as 'objects of transcendent beauty. They are the only places in the universe where Einstein's theory of general relativity shows its full power and glory.'[13]

But in spite of these exciting developments, particle physics seemed to have reached an impasse. Originally conceived to simplify phenomena and bring them under common theories, it was beginning to look as though whenever the nuclear top brass couldn't solve a particular equation, they invented a new particle – and then told their research minions to search for it. It was rather like looking down a telescope from the wrong end.

I took a road less travelled: a degree in theology at Fitzwilliam College, Cambridge, sitting at the feet of such luminaries as Donald Mackinnon, Simon Barrington-Ward and Charlie Moule – and went to India to lecture in physics.

State of the art

A brief account of physics has been given up to approximately 1930, when Einstein and Tagore encountered one another, and an even briefer cross section of where things stood by the mid-1960s has been indicated. The best state-of-the-art account of where particle physics stands now – or at least two or three years ago,

80 *Relativity and beyond*

because physics never stands still – is Roger Penrose's monumental *The Road to Reality*, all 1100 pages of it.[14] For reasons of brevity, we jump the first twenty-four chapters to the most concise summary that is possible of how things are at the beginning of the new millennium:

> Things have moved a great deal from these beginnings of an understanding of particle physics, as it stood in the first third of the 20th century. As we embark on the 21st century, a much more complete picture is to hand, known as the *standard model* of particle physics. This model appears to accommodate almost all of observed behaviour concerning the vast array of particles that are now known. The photon, electron, proton, positron, neutron, and neutrino have been joined by various other neutrinos, the muon, pions, ... kaons, lambda and sigma particles, and the famously predicted omega-minus particle. The antiproton was directly observed in 1955 and the antineutron, in 1956.
>
> There are new kinds of entity known as quarks, gluons, and W and Z bosons; there are vast hordes of particles whose existence is so fleeting that they are never directly observed, tending to be referred to merely as 'resonances'. The formalism of modern theory also demands transient entities called 'virtual' particles, and also quantities known as 'ghosts' that are even further removed from direct observability. There are bewildering numbers of proposed particles – as yet unobserved – that are predicted by certain theoretical models but are by no means implications of the general framework of accepted particle physics, namely 'X-bosons', 'axions', 'photinos', 'squarks', 'gluinos', 'magnetic monopoles', 'dilatons', etc. There is also the shadowy Higgs particle – still unobserved at the time of writing – whose existence, in some form or other (perhaps not as a single particle), is essential to present-day particle physics, where the related Higgs field is held responsible for the mass of every particle.[15]

Thus the number of particles has increased dramatically, but underlying them 'a more complete picture', known as the Standard Model, has emerged (see Figure 5.1). The diagram lacks the elegance of Rutherford's solar-system model of the atom, but there is a rudimentary simplicity about it. The two types of particle on the lower right of the diagram are named after Satyendra Nath Bose; the Higgs boson remains the Holy Grail of contemporary particle physics.

Along the road to Penrose's comprehensive summary there have been some losses and gains. Hoyle's steady state cosmology lost out to the Big Bang because it wasn't able to account for a uniform distribution of microwave radiation. It never posed any real threat to religious belief, and its rival leaves as many questions unanswered as it appears to solve. Einstein's cosmological constant is back in vogue again, though nobody understands why it seems to be so much smaller than quantum theory predicts.

In the 1970s it was proposed that weak particle interactions (e.g. β radioactive decay) are similar to electromagnetism (e.g. photons) except that they are mediated by heavy particles. This led to the electroweak theory, of which

Relativity and beyond 81

Elementary Particles

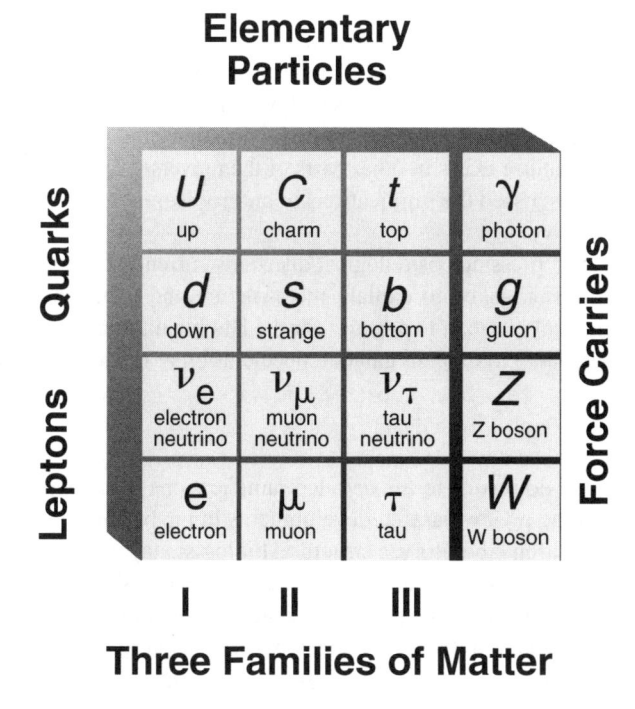

	I	II	III	
Quarks	U up	c charm	t top	γ photon
	d down	s strange	b bottom	g gluon
Leptons	ν_e electron neutrino	ν_μ muon neutrino	ν_τ tau neutrino	Z Z boson
	e electron	μ muon	τ tau	W W boson

Three Families of Matter / **Force Carriers**

Figure 5.1 The Standard Model of elementary particles. This is a considerable simplification of the much larger number of fundamental particles thought to exist by 1960s scientists, and subsequently.

the Glashow–Weinberg–Salam version is usually known as the Standard Model. Also in the 1970s Feltman and t'Hooft in the Netherlands proved that weak interactions can be described in terms of a class of theories that are renormalizable (i.e. the divergencies can be managed). This diverted attention from a hitherto useful mathematical device known as the S-matrix, which relates an initial to the final state of a system. Field theory – essentially Einstein's brainchild – became prominent again (this associates a field of oscillators at each point in space corresponding to every degree of freedom associated with a particle; quantum electrodynamics posits a field of such oscillators associated with electrons and photons).

Dark matter – matter that could comprise a large percentage of the mass of the universe but is undetectable except by its gravitational effects – was first proposed in the 1970s and is part of standard cosmology, though some physicists want to replace it with a tensor vector scalar theory which will alter the strength of gravity from place to place.[16] This sounds very *ad hoc*, but many physicists see dark matter the same way that junior researchers at Manchester University once viewed new particles overflowing from unsolved equations. According to a researcher at the University of Maryland, USA, 'If all the famous cosmologists defending the standard model had been deep frozen in the 1970s and then woken up today, not one of them would buy dark matter.'[17]

In 1987 Steven Weinberg proposed an anthropic explanation for the cosmological constant. (The anthropic principle argues that our universe is finely tuned to support life.) There are several versions of the anthropic principle, and it invites speculation as to whether or not life exists in other parts of the universe. Barrow, Polkinghorne and others have discussed the implications of anthropism in detail, and their arguments are persuasive.[18]

Among the most pressing cosmological questions currently exercising physicists are (1) dark matter invoked to explain the structure and behaviour of galaxies; (2) the cosmological constant to account for the Hubble diagram at high redshifts; and (3) 'quintessence' to explain galactic deceleration at much higher redshifts.

Conclusion

It is impossible to do justice to the detailed ramifications of modern physics in a few pages, let alone review parallel developments in the biological sciences. And yet today, in a research laboratory, a structural biologist may use techniques which embrace the full range of physical and biological sciences. We have therefore concentrated on those aspects of the physical sciences, especially particle physics and cosmology, which relate to the key themes in other chapters.

It has been important to evaluate Einstein's work in some detail, because it continues to be the best of its kind, and has enabled remarkable predictions to be made. There is no doubt in the mind of any competent scientist that space and time are not, as once imagined, independent variables. But when it comes to quantum theory, causation, and so on, there are significant divergences, and these have been noted.

Einstein's advice to philosophers and theologians not to see advances in physics as threatening or even, in most cases, significant has been underlined; we shall consider later how he understood the relationship between science and religion. He was deeply religious, and his beliefs were along the lines of what has been called 'cosmic religion', not entirely unlike Newton's 'thinking God's thoughts' – provided God never becomes too 'personal' (however theologians understand this term in relation to God).

A major weakness in the efforts of philosophers and theologians to relate to science has often been their desire to 'latch on' to scientific theories before they are firmly established. There was never any good reason for theologians to prefer Big Bang to steady state cosmology on account of superficial resemblances to early scriptural literature. (The same criticism may be levelled at some modern Hindus and Muslims, and the North American purveyors of Intelligent Design.)

It is therefore surprising that eminent theologians such as T. F. Torrance and Wolfhart Pannenberg have associated their thinking with specific scientific theories. Torrance uses the notion of space and time as a field serving as a medium for the interaction of God with the physical world and humanity.[19] Pannenberg relates space and time to 'the dynamics of the Spirit', and maintains that field theories 'claim priority for the whole over the parts',

> This is of theological significance because God has to be conceived as the unifying ground of the whole universe if God is to be conceived as creator and redeemer of the world. The field concept could be used in theology to make the effective presence of God in every single phenomenon intelligible.[20]

So far so good. But what are the implications of the fact that the field is regulated in structure by mass energy? And what if field theory had been discarded altogether in favour of the S-matrix?[21]

We give the final word to Einstein by reiterating his response to the question: what effect would relativity have on religion? 'None. Relativity is a purely scientific matter and has nothing to do with religion.'

6 Indian science comes of age

Prior to the arrival of the British, Indian science was uncoordinated and lacked an effective framework. But the introduction of higher education in English provided an institutional base for the development of science, technology and medicine, which were given pride of place from the outset. It introduced the youth of India to European scientific methodology – the 'Baconian philosophy' – stimulating them to high levels of achievement and giving them access to new and exciting fields of study and research. By the end of the nineteenth century the stage was set for the appearance of a group of remarkably gifted scientists whose work we shall consider in this chapter. Several of them shaped their research in accordance with their religious and philosophical beliefs; one collaborated extensively with Einstein, while others were inspired by his quest for unification among the sciences.

As we have seen, the nineteenth century in Europe was characterized by a trend whereby discrete branches of pure and applied science came ultimately to be understood as related to one another in novel ways. Heat, light, sound, electricity, magnetism, and all the familiar sectional headings of school textbooks were demonstrated to be capable of explanation by a few elegant theories. Electricity and magnetism merged into electromagnetic wave theories, and these were paralleled by similar discoveries in optics. The ether came to symbolize the supposed underlying stuff of the universe through which wave motions were propagated, and although the concept was ultimately discarded, it paved the way for Einstein's brilliant theories relating to gravitational and electromagnetic forces.

From the point of view of Indian scientists the progress of science in the West seemed to be the fulfilment of an important Hindu insight – the fundamental unity of all existence. Thus when Indian science became sufficiently advanced for basic research to be undertaken, some of the scientists chose to work in areas on the boundaries of established scientific disciplines. Jagadish Chandra Bose, for example, studied the phenomenon of pain in plants because he believed, as he put it, that 'In the multiplicity of phenomena, we should never miss their underlying unity.'[1]

Science becomes institutionalized

The growth of science was encouraged by the establishment of a number of universities and technical institutions throughout India. The Hindu College, which had been founded in 1817 by David Hare, an English watchmaker, later became Presidency College, now the premier institution in Calcutta University – and, according to *India Today*, the best college in India. Mathematics, astronomy and geography were given pre-eminence at the outset. In 1835 the Calcutta Medical College was set up. James Princep, head of the Calcutta mint and an able scholar, was so impressed by the standard of one chemistry class examined by him that he wrote:

> I do not think that in Europe any class of chemical pupils would be found capable of passing a better examination for the time they have attended lectures, nor indeed that an equal number... would be found so nearly on a par with their acquirements.[2]

The Elphinstone Institution had been founded in Bombay in 1827, later to become part of Bombay University. From the start its students were almost as rebellious as those at the Hindu College in Calcutta (though minus arch-rebel Henry Derozio!). At an early meeting of their Students' Literary and Scientific Society, one member denounced Hindu orthodoxy in the name of Western knowledge.[3] It is interesting that they chose a scientific meeting to do this.

Applied science appears to have got off to a slow start, in spite of the British East India Company's need for qualified technologists. But in 1847 the Roorkee Engineering College began to attract extremely capable students. In 1858 the Poona College of Engineering came into being, and an industrial training school in the Madras Presidency became a fully fledged college of engineering. They were affiliated to Bombay and Madras universities.

Calcutta, Bombay and Madras universities were modelled upon the London system and constituted as affiliating and examining bodies; government, missionary and private colleges were responsible for tuition. All three were initially constituted with four faculties – arts plus science, law, medicine and engineering; in 1902 Calcutta introduced a BSc degree. In Bombay and Madras presidencies several colleges rapidly gained pre-eminence, notably Elphinstone, Wilson, St Xavier's and the Grant Medical College in Bombay, and Madras Christian College and several Jesuit colleges in Madras. Vernacular education as well as English was encouraged in the Bombay Presidency, and Tamil-speaking South Indians were at pains to demonstrate the adequacy of Tamil for scientific purposes. In 1916 a Tamil Scientific Terms Society was started in Madras. Osmania University in Hyderabad used Urdu as the medium of instruction for modern sciences at a high academic level.[4]

In the northwest provinces the government approved the establishment of Lahore University in 1869. Its objectives included: 'The diffusion of European science, as far as possible, through the medium of vernacular languages of the Punjab'.[5] Edwardes College in Peshawar was founded in 1901 and came initially under the aegis of Lahore University. It became part of Peshawar University in the 1970s, and is now the most sought-after independent higher education institution in Pakistan.

The Indian Institute of Science (IISc) in Bangalore was the first of what became the national technological institutes, some of which are currently ranked near the top of the global tables of achievement. It was founded in 1911 under the auspices of Mysore State following a substantial endowment by J. N. Tata, the Parsee industrialist. In recent years the IISc has become a haven for brahmins who cannot gain admission to the universities of Tamil Nadu on account of their high admission quotas for non-brahmins.

St Stephen's College in Delhi was founded in 1881. Delhi University was established in 1922, first under the aegis of Calcutta, then Lahore universities for degree purposes. In 1940 St Stephen's moved from Kashmere Gate to the main university campus. Though initially geared to liberal arts subjects it recently began to offer honours physics and chemistry (though not biology).

Scientific institutions were caught up in the tensions generated by the growth of national feeling against British rule. Thus there was an outcry when the Government withheld funds from the Calcutta University College of Science, giving them instead to the Bangalore IISc and the Bombay Royal Institute of Science, both 'staffed, managed and controlled entirely by the British element'.[6] But such official discrimination, far from intimidating the scientific community, had the opposite effect, with the result that by 1920 several outstanding southerners, including Nobel Prize-winning C. V. Raman, deserted the south for Calcutta. Lord Curzon's attempts to restructure the universities met with the same level of opposition as practically everything else that he did. But by then Indian science had come of age, and its practitioners represented the first truly indigenous phase of scientific development. As P. C. Roy put it:

> Following the inoculation of the Hindu mind with western ideas at least two generations had to elapse before any tangible result could be achieved in the shape of that steadfast devotion to physical science which alone can bring forth originality.[7]

By the early years of the twentieth century science was occupying pride of place in educational institutions throughout India. It was mainly, though not exclusively, taught in English, though several other languages such as Tamil were being developed to incorporate new scientific concepts.

The rediscovery of ancient science

It was at the beginning of the twentieth century that Indian scientists and other scholars began to rediscover the ancient Indian science to which we have alluded. Their discoveries, though not always soundly based, were much more plausible than those of Dayanand Sarasvati, who tended to read modern science and technology back into the Vedas. It is also important to recognize that science had not quite achieved the breakthroughs which occurred between 1905 and 1915 with Einstein's remarkable theories – though some Indian scientists were among the first to appreciate the significance of these and to collaborate with Einstein in his further work.

The scholarly rediscovery of ancient Indian science was set out in several substantial tomes by Prafulla Chandra Roy and Brajendranath Seal. Roy published his *History of Hindu Chemistry* in two parts at the turn of the century, and Seal published *The Positive Sciences of the Ancient Hindus* soon afterwards. Roy was a professor of chemistry, and Seal was a philosopher who occupied the Chair of Mental and Moral Science at Calcutta University from 1913–20, when he became Vice-Chancellor of Mysore University. Roy paid tribute to his 'encyclopaedic knowledge', and the two of them were closely associated.

Throughout the nineteenth century, first European and later Indian historians had taken a keen interest in India's past, and the careful study of ancient coins, seals and inscriptions opened up a vast quantity of new knowledge which made it possible for Seal and Roy to compile their scientific histories. Roy states that he was also influenced in his decision to undertake the *History of Hindu Chemistry* through reading Kopp's *Geschichte der Chemie*. In his preface Roy explains:

> We are now in possession of ample facts and data, which enable us to form some idea of the knowledge of the Hindus of old in the fields of Philosophy and Mathematics including Astronomy, Arithmetic, Algebra, Trigonometry and Geometry. Even Medicine has received some share of attention.... One branch has, however, up till this time, remained entirely neglected, namely, Chemistry.[8]

Seal's book, *The Positive Sciences of the Ancient Hindus*, was even more comprehensive. In it he attempted to reinterpret the classical Sāṃkhya system in accordance with Western evolutionary concepts:

> The world evolves out of Maya, so that Maya in the Vedanta replaces the Prakriti of the Samkhya. But Maya, and by implication the world, originate out of Brahma, not by a process of evolution, but of Vivarta (self-alienation). The self-alienation of the Absolute, acting through Maya, produces in the beginning Akasa – one, infinite, ubiquitous, imponderable, inert, and all pervasive. The world thus begun goes on evolving in increasing complexity.[9]

The concept of evolution in Seal's works corresponds to the Sanskrit term *pariṇāma* which certainly can carry the idea of change and development. But *pariṇāma* in classical Sanskrit is not what evolution meant for Darwin or even Spencer. Seal translated *ākāśa* as ether, which was reasonable enough until Einstein dispensed with the notion. He may well have been instrumental in encouraging Jagadish Bose to investigate the response of plants to external stimuli: 'The Hindu Scriptures teach that plants have a sort of dormant or latent consciousness, and are capable of pleasure and pain.'[10]

Neither Roy nor Seal would have been able to collect and translate so much information about India's scientific past had it not been for the earlier work of European Sanskritists like Sir William Jones and Max Müller. Their scientific histories, though historically open to criticism, were a timely reminder to their generation of the achievements of India's past. But it is doubtful whether, in the absence of growing political nationalism, either of the two works which have been mentioned would have been so widely read.

Some of the distinguished scientists of this period will be considered in more detail, noting in particular their fascination for the boundary areas between the sciences, and their interest in Einstein's quest for a unified field theory.

Eminent scientists

Prafulla Chandra Roy

G. K. Gokhale, the prominent nationalist and reformer, once declared: 'Where will you find another scientist in all India to place by the side of Dr J. C. Bose, or Dr P. C. Roy?'[11] Viewed from the standpoint of national politics, Jagadish Chandra Bose and P. C. Roy were the most important members of a group of outstanding scientists who dominated Calcutta's lecture rooms and laboratories during the first two decades of the last century. It is not surprising, therefore, to find that more has been written about them than about their fellow-scientists. But it does not necessarily follow that they were the most able scientists of their generation, and it was ultimately C. V. Raman who gained the greatest degree of international recognition with the award of a Nobel Prize in 1930.

P. C. Roy (1861–1944) moved to Calcutta from Jessore with his family in 1870 where he attended the Hare and Albert schools. He came under the influence of two teachers who were members of the Brahmo Samaj. He records: 'The tenets of the Brahmo Samaj were explained to us – how it differed from other forms of faith in that it was not based on revelation, but had to draw more or less on rationalism and intuition.'[12] Roy's parents wanted to send him to the prestigious Presidency College, but could not afford the high fees, so they sent him instead to Pandit Iswarchandra Vidyasagar's Metropolitan Institute, where he developed a keen interest in chemistry. He won a scholarship to Edinburgh, and in 1881 was appointed to a lectureship at Presidency College, eventually becoming head of the chemistry department.

At several points in his autobiography and published research papers Roy refers to intuition in relation to scientific research. The account of his own discovery of mercurous nitrite is a good illustration of what he meant by it:

> Having recently had occasion to prepare mercurous nitrate in quantity by the action of dilute [nitric] acid in the cold on mercury, I was rather struck by the appearance of a yellow crystalline deposit. At first sight it was taken to be a basic salt, but the formation of such a salt in a strongly acid solution was contrary to ordinary experience. A preliminary test proved it, however, to be at once a mercurous salt as well as a nitrite.[13]

Roy was not the first to notice the yellow crystalline deposit which appeared during the preparation of the stable nitrate, but he was the first person to reflect upon the likelihood of an impurity consisting of a basic salt remaining intact in the presence of a strong acid. He therefore tested the impurity and discovered that he had prepared mercurous nitrite.

This 'intuitive' discovery took place in 1895, and it was immediately recognized as being of considerable importance. Roy followed it up with the preparation of a whole series of new compounds:

> As one new compound followed the wake of another, I took up their examination with unabated zeal. In short, I could fully enter into the feelings of one of the illustrious makers of modern chemistry, the immortal Scheele – 'There is no delight like that which springs from a discovery; it is a joy that gladdens the heart.'[14]

In 1912 Roy attended the Empire Universities Congress and read a paper on the vapour density of ammonium nitrite – which was something of a surprise to the delegates since ammonium nitrite was not supposed to exist in a sufficiently pure state for such a determination to be possible. As a professor at Presidency College, Roy exerted a strong influence over his colleagues and students, and gathered around himself a team of exceptionally brilliant researchers. He records:

> The year 1909 opened a new chapter in the history of chemical research in Bengal. In that memorable year some members of the brilliant group of students who were afterwards destined to play a conspicuous part in notable research took their admission in Presidency College.... The bonds existing between them and me were as subtle as those of chemical affinity.[15]

Roy was concerned not only that his colleagues and students should do good scientific research, but that their achievements should be recognized as those of Indians who had done their work in India:

> Taking into account the physical sciences it may be mentioned that Professor Raman of 'Raman Effect' is practically self-taught and all his brilliant researches have been carried on in the laboratories of Calcutta.... Ghosh and Saha did not care to have a D.Sc. (Lond.) suffixed to their names for fear they would be thereby lowering the doctorate of their own alma mater. Satyendra Nath Bose (of Bose–Einstein statistics fame) although he went abroad for a time to rub shoulders with eminent mathematicians and physicists, also from the same motives, fought shy of a foreign degree.[16]

Roy's nationalistic inclinations also led him to be concerned with the economic and commercial implications of the preparation of chemicals, and he investigated ways in which chemicals could be prepared cheaply in India instead of being imported. He made sodium phosphate – a salt widely used as a fertilizer – from bone ash and sulphuric acid, and in the course of a public lecture outraged some orthodox Hindus by inviting them to put some of the burnt bone ash into their mouths to stress that it was merely a chemical compound. The exact date of the incident is not recorded by Roy, but from the context it may be assumed to have taken place between 1890 and 1900, which suggests that many people were still uneasy about defilement associated with touching the dead in spite of more than half a century of exposure to Western medical and surgical procedures.[17]

Roy prepared drugs, some of them Ayurvedic, and exhibited them at the Indian Medical Congress in Calcutta in 1898. But many people were averse to his *deshi* (local) products, and considered them inferior to imported items. His enthusiasm for indigenously prepared chemicals led to the establishment in 1892 of the Bengal Chemical and Pharmaceutical Works. He was convinced that if more chemistry graduates would pay attention to the practical applications of their subject instead of opting for overcrowded government and mercantile jobs, then unemployment among graduates could be reduced.

With regard to his religious convictions, Roy maintained a firm belief in 'Providence' throughout his life. Reason and intuition, the twin principles he had derived from the Brahmo Samaj, excluded a large part of popular Hinduism, and superstitious practices such as not touching dead bodies and bone ash derived from them were irrational and therefore not to be condoned. But the idea of a single omnipresent being whose guiding hand shapes human destinies is a belief which frequently recurs in Roy's autobiography:

> Whatever field I have ploughed I have ploughed as a humble instrument in the hand of Providence; my failures are my own; to err is human. But my successes, if any, are to be attributed to the guidance of the All-knowing, who chose me to be His humble instrument. After all, a Divinity shapes our ends.[18]

Elsewhere he acknowledges the inspiration of a number of Christian writers:

> In my maturer years Martineau's *Endeavours after the Christian Life* and *Hours of Thought*, Theodore Parker's and Channing's writings have been my favourite companions.... All through my varied activities I felt the force of the saying 'I commit myself to Thee, O Lord! Make me Thy agent.'[19]

The quantity of detailed material contained in Roy's autobiography has made it possible to give a reasonably comprehensive account of his beliefs, and it is clear that he saw his main contribution to the national cause as being the degree of excellence of his own scientific research. He also believed very strongly that reason and intuition play an important role in scientific discovery, a view shared with others such as Jagadish Chandra Bose.

In one important respect, however, Roy was unlike other contemporaries, namely, his willingness to work within the framework of a single scientific discipline, that is, chemistry. Many of his scientific colleagues deliberately concentrated their efforts upon the boundary regions between established fields of scientific research, and this was most conspicuously true of his fellow nationalist 'hero', Jagadish Chandra Bose.

Jagadish Chandra Bose

On 12 November 1896, Lord Hamilton wrote to Lord Elgin as follows:

> There is a strong feeling here that the Government should in some way mark its appreciation of Dr. J. C. Bose's remarkable labour and researches in science. The highest scientists here express great admiration of the little man, who is undoubtedly the foremost scientific authority amongst the educational officers of the Indian Government, both European and Native.[20]

The previous year the 'little man' had achieved international fame with a paper entitled 'The Polarization of Electric Waves' which was read to the Asiatic Society. Not only was the paper a capable piece of theoretical research, but it was subsequently reproduced as a lecture and accompanied by an experimental demonstration.

Jagadish Chandra Bose (1858–1937) was born into a *kayastha* family in Bikrampur in what is now Bangladesh (near Dacca). Ashis Nandy describes this eastern region of Bengal as 'the backyard of *babu* culture – a damp, marshy, dialect-speaking nest of provincials where ugly *Bangal* ducklings dreamt of becoming elegant Calcuttan swans.'[21]

Bose's father was an active member of the Brahmo Samaj whose 'religious enthusiasm was associated with a lively interest in science and technology, an amalgam of interests that was to become a formal synthesis in the son much later'.[22] Bose developed an early fascination for plant and animal life, but after admission to St Xavier's College, Calcutta, discovered that the curriculum obliged him to do physics rather than botany or zoology (the same would be true today at St Stephen's College, Delhi). Fortunately, the quality of physics teaching was such that he temporarily abandoned his love for the biological sciences, recovering it to some extent in London University, where he studied medicine. He became ill through working with certain solvents, and went to Cambridge (Christ's College) in 1882, where he read chemistry, physics and botany for the natural sciences tripos. Lord Rayleigh, Maxwell's successor as Cavendish professor of physics, made a deep impression on him.[23] In 1895 Bose was appointed professor of physics at Presidency College, Calcutta.

To appreciate the significance of Bose's work on the polarization of electromagnetic waves, it is necessary to refer back to Maxwell's and Faraday's earlier research. At the end of the eighteenth century Faraday had demonstrated that a coil carrying electricity can induce a current in another coil placed some distance from it. But how did the energy cross the gap between the two coils? Or, for that matter, how did light energy manage to cross vast areas of empty space? Maxwell postulated the ether, and argued that electrical, magnetic and optical energy is transmitted through this medium in the form of three-dimensional waves.

Maxwell's electromagnetic waves effectively brought together three branches of physics, and possibly this aspect of his work was what initially fascinated Bose. The idea of an all-pervading ether was later discarded, but Maxwell's fundamental concept of a three-dimensional wave which can be represented mathematically by two vectors at right angles moving with the velocity of light in a direction perpendicular to the plane containing them remains valid. Bose's experiments consisted of demonstrating that it is possible to eliminate one or both of Maxwell's electromagnetic vectors by passing radio waves through a crystal called nemalite.

From this brief summary of the research for which Bose achieved international fame it will be clear that he had chosen to work in an area where the underlying unity and interrelatedness of the sciences might seem reasonably apparent. But he was not content to accept the limitations imposed by remaining within the physical sciences. He therefore reverted to his earlier concerns when he was admitted to Christ's College, Cambridge. Combining this with his later work he began to study parallels between the behaviour of metal 'coherers' and plant and animal responses to external stimuli. This led him into regions where the boundary lines were even less distinct:

> In the pursuit of my investigation, I was unconsciously led into the border regions of physics and physiology and was amazed to find boundary lines vanishing and points of contact emerge between the realms of the Living and the Non-living.... A universal reaction seemed to bring together metal, plant and animal under a common law.[24]

Elsewhere he refers to the continuity of life processes between the living and the inanimate:

> There is no break in the life-processes which characterize both the animate and the inanimate world. It is difficult to draw a line between these two aspects of life. It is of course possible to delineate a number of imaginary differences, as it is possible to find out similarities in terms of certain other general criteria. The latter approach is justified by the natural tendency of science towards seeking unity in diversity.[25]

Bose's research career can be divided chronologically into three phases. From 1894 to 1899 he studied the properties of electromagnetic waves; from 1899 until 1902 he compared physical responses in the living and the non-living; and from 1903 until 1933 (a few years before his death) he concentrated on response phenomena in plants, which he regarded as intermediate between those of animals and inanimate matter. Towards the end of his life be became increasingly authoritarian, and the reliability of some of his data was questioned.

On the occasion of the inauguration of the Bose Institute in Calcutta in 1917, Bose set his quest for a wider synthesis between the sciences against the background of what he described as the excessive specialization of scientists in the West, and claimed that India was well suited by her past to provide a corrective:

> The excessive specialization in modern science has led to the danger of losing sight of the fundamental fact that there can be one truth, one science which includes all branches of knowledge. How chaotic appear the happenings in

> Nature! Is nature a cosmos in which [the] human mind is some day to realize the uniform march of sequence, order and law? India through her habit of mind is peculiarly fitted to realize the idea of unity, and to see in the phenomenal world an orderly universe. It was this trend of thought that led me unconsciously to the dividing frontiers of the different sciences and shaped the courses of my work in its constant alteration between the theoretical and the practical, from the investigation of the organic world to that of organized life and its multifarious activities of growth, of movement, and even of sensation. Thus the lines of physics, of physiology, and of psychology converge and meet.[26]

Bose's interest in botany and physiology was in part a reversion to an early childhood passion, and there was also an element of reaction against the West's emphasis upon specialization. But he believed that by concentrating his attention upon the boundary areas between the different physical and biological sciences he would ultimately help to demonstrate the underlying unity of all things. The scientist's quest was thus virtually a religious activity, the ultimate goal of which was the discovery of unity in diversity. The following two quotations from Bose's speeches and writings, taken together, describe the 'theological bias' of the scientist in the desire to question and understand, and the 'moment of truth'.

> In my scientific research... an unconscious theological bias was also present.... It is forgotten that He, who surrounded us with this ever-evolving mystery of creation, the ineffable wonder that lies hidden in the microcosm of the dust particle, enclosing within the intricacies of its atomic form all the mystery of the cosmos, had also implanted in us the desire to question and understand.[27]

Bose's language was strongly coloured by Sāṃkhya concepts – 'microcosm' and 'ever-evolving mystery of creation', for example. Subrata Dasgupta elaborates on this in his biography of Bose.[28]

The ultimate aim of scientific discovery, Bose argues, is the realization of the one in the all:

> When I came upon the mute witness of these self-made records and perceived in them one phase of a pervading unity that bears within it all things: the mote that quivers in ripples of life, the teeming life upon our earth, and the radiant suns that shine above us – it was then that I understood for the first time a little of that message proclaimed by my ancestors on the banks of the Ganges thirty centuries ago. 'They who see but one in all the changing manifestations of this universe, unto them belongs Eternal Truth – unto none else.'[29]

Thus the work of the scientist, properly conducted, is a religious quest whereby we are drawn naturally to search for the wonder that is at the heart of all existence.

A corollary of this conviction is that if scientists wish to make any significant progress, they should approach their work with a degree of sensitivity and reverence which was not particularly characteristic of the sort of techniques sometimes advocated by Western scientists at the end of the nineteenth century. Bose believed that the true scientist must learn to evoke, look and listen, rather than probe and analyse from a distance. But although he disliked what he sensed to be the aggressive Western attitude to science, his methods were acceptably Western, and the results of his early botanical work were acclaimed in Europe and America. Like Roy he believed in the importance of intuition, and his attempts to prove the existence of consciousness in lower forms of life were essentially intuitive.

Bose's biological researches were prompted initially by the discovery that an electric wave receiver seems to show signs of fatigue after continued use. He began to wonder how far it was legitimate to compare the responses of living and inert matter to external stimuli, and whether or not plants could be shown to possess some sort of latent consciousness. But he was unable to progress very far, and the evaluation of his work in the 1945 edition of the *Encyclopaedia Britannica* – eight years after his death – was a fair one: his research 'was so much in advance of his time that precise evaluation is not possible'.

Ashis Nandy – a veteran Gandhian and a perceptive critic of Eurocentrism – compliments Bose for being able to straddle both the 'cosmic unity' of Brahmoism, combining the 'pure monism of Vedānta' with 'a conspicuous monotheistic element that showed the influence of Christianity and Islam', and the 'projective system of the little cultures of India'. He continues,

> Here the abstract, universalist nondualism of the greater Sanskritic culture became a pan-psychic tendency to see the world as a living organism where the natural entities were not only endowed with life, but with the ability to manipulate human behaviour and fate. Bose's friends Vivekananda and Nivedita understood this much better than did the Brahmo leadership. Ideologically the former were more passionately committed to the philosophy of Vedānta, but in practice they constantly invoked the little culture's more dualist anthropomorphism. Understandably, their idiom reached beyond the perimeters of urban, westernized, upper-caste Bengal whereas Brahmoism slowly drowned in the blue blood of its followers.[30]

Nandy's concluding caustic observation would have been more true at the tail end of the nineteenth century than in the years that followed, during which Rabindranath Tagore assumed leadership of the Brahmo Samaj. Tagore viewed Bose's researches as support for his own universalism:

> European science is following the way of our philosophy. This is the way of unity. One of the major obstacles which science has faced in forging this unity of experience is the differences between the living and the non-living. Even after detailed research and experimentation scholars like Huxley could not transcend this barrier. Venturing this excuse biology has been maintaining

Satyendra Nath Bose

Satyendra Nath Bose (1894–1974) was the eldest child of an employee in the engineering department of the Calcutta railway company. When he was three years old an astrologer credited him with exceptional intelligence and predicted that he would achieve great fame. At high school in Calcutta he set a new record in mathematics by scoring 110 marks for a maximum of 100 (he solved one question in two different ways). He read physics at Presidency College, and in 1916 was appointed a lecturer there, moving to Dacca University in 1921, where a research paper by him about light quanta attracted the attention of Albert Einstein.

To appreciate Bose's work it is necessary to trace the development of ideas about heat from the beginning of the nineteenth century when Joule first realized that it is not a substance but a form of energy. Attempts were made to describe this energy in terms of small, elastic molecules obeying normal statistical laws. At the same time it was also known that under certain conditions light possesses particle properties similar to those of heat molecules, so it was only natural to apply thermodynamic statistical laws to optical phenomena. But the answers came out wrong, and Bose was the first person to realize why.

We noted earlier the reasons why the possible permutations for particles vary according to certain conditions such as whether or not they are identical. The types of statistics corresponding to these situations are Maxwell–Boltzmann, Bose–Einstein and Fermi–Dirac statistics. Heat molecules may be distinguished from one another and may crowd together, and therefore obey the first type of statistics. Photons, that is, light particles, are indistinguishable and interchangeable but are not subject to the Pauli Principle, and hence obey Bose–Einstein statistics. And electrons not only cannot be distinguished from one another, but are subject to crowding restrictions and hence obey Fermi–Dirac statistics.

Clearly, Bose's discovery was an important one, but he did not continue with the same line of research for long, and after a brief period of research with Madame Curie in Paris in 1924, he joined Einstein in his quest for a unified field theory. In 1953 he managed to solve some of the wave equations governing the relationship between electromagnetism and gravitation, but by then the scientific world had lost interest in wave theories, and little attention was given to Bose's later research.

Bose's name is associated with two of the most vital areas of contemporary physics. The first concerns the phenomenon of superconductivity, and in particular the properties of helium, which change just above absolute zero. The particle named after Bose, the boson, refers to a class of particles which obey Bose–Einstein statistics. The Higgs boson has yet to be discovered experimentally. Although Bose died three decades ago, his legacy is more persistent than that of almost any other scientist – save legendary figures such as Newton and Einstein.

Srinivasa Ramanujan

Unlike the previous Indian scientists Srinivasa Ramanujan (1887–1920) was born into a poor Tamil family in South India. He died in his home village at the age of thirty-two of tuberculosis after a short but brilliant career which took him to Cambridge where he worked with G. H. Hardy and others on a range of crucial mathematical topics. The cross-cultural friendship between Hardy and Ramanujan is the subject of a new film by Dev Benegal, starring Stephen Fry, best known for his roles in *Blackadder* and as P. G. Wodehouse's Jeeves.

Ramanujan's mathematical gifts were apparent early in life. By the age of thirteen he could calculate $\sqrt{2}$, π and e to any number of decimal places. He derived Leonhard Euler's equation and several others unaware of the fact that they had already been discovered in Europe. In Cambridge Ramanujan's most original research was in the area of fractional differentiation, not, as is often supposed, on the theory of numbers. He also worked on hypergeometric series, partitions, definite and elliptical integrals, highly composite numbers and number theory. In spite of G. H. Hardy's famous toast, 'To pure mathematics; and may it remain useless forever,' much of Ramanujan's theoretical work has been of enormous practical scientific value.

Ramanujan maintained that his mathematical researches were guided by his religious beliefs. In the absence of published details by him it is difficult to assess this claim. However, P. C. Mahalanobis commented on the religious implications of his research as follows:

> He was eager to work out a theory of reality which would be based on the fundamental concepts of 'zero', 'infinity' and the set of finite numbers. He sometimes spoke of 'zero' as the symbol of the absolute (*Nirguna Brahman*) of the extreme monistic school of Hindu philosophy, that is, the reality to which no qualities can be attributed, which cannot be defined or described by words and is completely beyond the reach of the human mind; according to Ramanujan, the appropriate symbol was the number 'zero', which is the absolute negation of all attributes. He looked on the number 'infinity' as the totality of all possibilities which was capable of becoming manifest in reality and which was inexhaustible. According to Ramanujan, the product of infinity and zero would supply the whole set of finite numbers. Each act of creation... could be symbolised as a particular product of infinity and zero, and from each such product would emerge a particular individual of which [the] appropriate symbol was a particular finite number.[32]

Thus zero is the ultimate integrative symbol, because it absorbs every number it comes in contact with and even infinity. It is also the null point of a system of coordinates.

Ramanujan's legacy lives on. At one end of the spectrum there is the recent film about his friendship with Hardy. More academically, the 2005 Mordell lecture at Cambridge University was given by a mathematician from Princeton about 'Ramanujan's conjecture', which is a summation equation in analytic number theory. This, the lecturer explained to a large audience, has implications for Selberg's eigenvalue conjecture, Frobenius' eigenvalues and Clozel's theorem; it also relates to the eigenformations of Hecke algebra, the infinite dimensions of Hilbert space, analytic homomorphism and ergodic theory. *Caveat lector*!

Nehru saw in Ramanujan's short career a microcosm of India's destiny:

> Ramanujan's brief life and death are symbolic of conditions in India. Of our millions how few get any education at all.... If life opened its gates to them... how many among these millions would be eminent scientists, educationists, technicians, industrialists, writers and artists, helping to build a new India and a new world?[33]

Other eminent scientists

Meghnad Saha (1893–1956) was interested in the work of Bose and Einstein, and published an early paper on aspects of the Theory of Relativity. But his main contribution to science was in astrophysics, and he was responsible for the idea that in a very hot star electrons may be stripped from their parent nucleus to form an electronic 'gas'.

At about the same time that Einstein published his General Theory of Relativity, Bohr and Rutherford succeeded in proving that atoms can be thought of as miniature planetary systems of electrons surrounding a parent nucleus. Under normal circumstances electrons are subject to the force field of the nucleus, and on account of the Pauli Principle obey Fermi–Dirac statistics. Saha proved that in very hot stars electrons are ionized and removed so completely from their nuclei that they obey Maxwell–Boltzmann statistics and can be treated like ordinary molecules. His later work included the application of his stellar theories to the behaviour of molecules in the upper atmosphere where intense ultra-violet radiation produces a similar ionization effect.

Why did Meghnad Saha not share the religious and philosophical beliefs of Roy, the two Boses and Ramanujan? His youngest daughter, Sanghamitra Roy, currently teaching history at Dhaulat Ram College in Delhi University, had this to say:

> My father believed that the Vedic scriptures are man-made. He came from a poor *vaiśya* family, and was educated in a school run by brahmins who treated him as an outcaste. He said to me 'brahmins practise untouchability'. However he recognized that not all brahmins are like that and it was a brahmin, Sir Asutosh Mukherji, who did help to fund his education. He once said: 'Blacksmiths, leather workers and other *chamars* are not treated well and yet they are the iron pillars of our society.'
>
> Meghnad was very interested in ancient India and knew Sanskrit well; in fact when he married he corrected the grammar of the priest's *ślokas*. He was interested in Buddhism and admired its moral and social system – in fact my own name is that of Ashoka's first woman emissary to Sri Lanka. He knew

Tagore personally and they shared many things such as opposition to caste discrimination, but he did not agree with Tagore's beliefs in God.

My father was a leftist and admired Stalin's five-year plans. He advised Nehru who sought his advice, but they fought over many things. Nehru listened more to Homi Bhabha. My father was a founder member of the University Grants Commission and involved in our first Five-Year Plan. He also won a seat in the Lok Sabha from Bengal. He was a friend of Netaji Subhas Chandra Bose, who wanted science to play a part in national politics.

My father was the main architect of a plan in the 1950s to harness the water in India's rivers more productively – the Riverene scheme. This was devised to control the extensive floods in Bengal and was based on small dams, which would have been very different from Nehru's huge ones based on western models such as the Bhakra Nangal Dam.

Meghnad was very interested in European history and knew German.[34]

Chandrasekhara Venkata Raman (1888–1970) was the first Asian to win the Nobel Prize in physics. He was born in Tiruchirapalli in Tamil Nadu and developed a love for physics and music at an early age from his schoolteacher father. He graduated in physics from Presidency College in Madras University with the highest marks on record, and in 1915 was appointed professor of physics at the new Science College in Calcutta University.

Raman's early research in acoustics was eventually broadened into the study of the more general properties of waves. In 1921, while representing Calcutta University at the Congress of the Universities of the British Empire in London, he visited St Paul's Cathedral, where he noticed that a whisper could be heard clearly in two places under the dome. This discovery led him to study comparable non-acoustic wave properties.

In 1928 'C. V.' announced the discovery of the phenomenon named after him, the Raman effect. This is a scattering process whereby light reflected from molecules in a substance differs in wavelength from the incident light (usually the waves become longer). It is important to distinguish the Raman effect from fluorescence in that it is not a resonance phenomenon – the incident light is not the same as the absorption band of the substance, and the scattered light is much weaker than fluorescent light.

Raman was able to prove that the changes in wavelength associated with scattered light in a medium correspond to energy-level differences in the substance associated with molecular properties such as vibration and rotation. Raman spectroscopy, especially when used with laser light, has proved a sensitive tool for exploring the detailed properties of molecules.

Raman's discovery followed fairly logically from Bose's and Einstein's work on photons. Smekel had theoretically predicted a change in wavelength corresponding to the Raman effect in 1922. The discovery brought Raman honours from all over the world. He joined the Indian Institute of Science in Bangalore in 1933, and the following year established the Indian Academy of Sciences. A gift of land from the Mysore State government enabled him to set up the Raman Research Institute, which he surrounded with a large garden full of eucalyptus trees: 'A Hindu is required to go to the forest in old age, but instead of going to the forest, I made the forest come to me', he remarked.[35] He regarded his scientific work as a religious activity:

> Research is a strange work. Success in research brings limitless joy whereas failure pushes one to deep despair. New discoveries confirm the existence of God, and... we have to find Him in the Universe.[36]

In 1968, two years before his death, C. V. Raman put forward a new theory of the physiology of vision. The anniversary of the discovery of the Raman effect, 28 February, is celebrated as National Science Day in India.

Subrahmanyan Chandrasekhar (1910–95), the son of Tamil parents, was born in Lahore. His father was the brother of C. V. Raman. In 1918 the family moved to Madras, where at the early age of fifteen he joined Presidency College to study science. He was inspired by Ramanujan's career and wanted to read mathematics, but his father considered physics a better choice for a civil service job, so he began to specialize in physics.

The civil service did not appeal to the young physicist, who attracted the attention of Meghnad Saha (then at Allahabad) with a research paper on interactions between photons and charged particles (the Compton effect). He accepted a scholarship to go to England in 1930; he moved between Cambridge, Germany and Denmark (the Bohr Institute in Copenhagen).

Chandrasekhar began his research career in earnest by studying 'white dwarf' stars. These have the mass of an ordinary star like the sun, but they are much older and have run out of nuclear energy. They have therefore collapsed to a size more like that of the earth, giving them an enormous density. He realized that white dwarf physics must be relativistic.

Chandrasekhar also considered why it is possible for these stellar objects to maintain their small earth-like size while not collapsing even further as a result of their enormous innate gravity. His explanation was based on Pauli's principle that no two electrons can be close enough to occupy the same quantum state; this generates a special electronic pressure which balances the gravitational force. But this will only happen if the original mass of the star is less than about 1.4 times the mass of the sun – later called the 'Chandrasekhar effect'. If it is larger, then the gravitational pull of the collapsed star will be so huge that the electronic pressure will not be sufficient to prevent even further collapse to produce a neutron star or even a black hole.

'Chandra' made many good friends in Europe, but clashed badly with Arthur Eddington, who accused him of 'stellar buffoonery'. Eddington was wrong, but nobody dared tell him so. In 1983 Chandrasekhar published a determinative book on black holes, the same year that he was awarded the Nobel Prize. By this time he was settled at the University of Chicago. Sir Brian Pippard, formerly Cavendish professor and President of Clare Hall, Cambridge, recounts that he once visited him there with a mathematical problem which he was finding difficult.

Chandra wrote it up on a blackboard, paused for a moment, and then said: 'Yes, it is difficult; almost difficult enough for me to solve', which he then did.

Chandrasekhar described himself as a 'lonely wanderer in the byways of science'. He liked to compare his research to the climbing of mountains:

> The pursuit of science has often been compared to the scaling of mountains, high and not so high. But who among us can hope, even in imagination, to scale the Everest and reach its summit when the sky is blue and the air is still, and in the stillness of the air survey the entire Himalayan range in the dazzling white of the snow stretching to infinity? None of us can hope for a comparable vision of nature and the universe around us, but there is nothing mean or lowly in standing in the valley below and waiting for the sun to rise over Kanchenjunga.[37]

Most of the scientists so far mentioned were religious, with the exception of Meghnad Saha. Homi J. Bhabha (1909–66), the architect of India's nuclear programme, was a Parsi, and Abdul Kalam, President of India, is a Muslim, though at pains to stress aspects of spirituality common to all religions. Several were quite explicit in their belief that religion, or, at least, philosophy based on religion, is consonant with science, though most would have been too modest in this respect to echo the sentiments of Sir Asutosh Mookerjee, Vice-Chancellor of Calcutta University for many years, that: 'Science in its ultimate essentials echoes the voice of the living God.'[38]

Conclusion

Following the decision to introduce higher education through the medium of English, universities were established in Calcutta, Bombay and Madras presidencies. These complemented existing institutions, such as the Hindu College (later Presidency College), and were based on the London system, which meant that these colleges could be assimilated and supplemented. From the start science and medicine were given pride of place, though technological centres such as the Indian Institute of Science in Bangalore soon followed.

By the end of the nineteenth century Indian science was well established, and the scientists themselves were making great strides in their research which they saw as a counterbalance to the achievements of European science. They interpreted their work increasingly as a contribution to the nascent national movement, and also as the expression of aspects of their religious and philosophical traditions – the two went together. But their researches were also hailed by the international scientific community, and some of them eventually collaborated with the best minds in the West.

Several of these scientists, in addition to their research, set about trying to rediscover ancient Indian science. Their findings tended to be more comprehensive than those of the Reformers, though occasionally marred by the limitations of the science of their day – thus Brajendranath Seal, P. C. Roy and others could hardly be blamed for comparing the notion of *ākāśa* (space) to the ether wind a few years before Einstein abandoned the concept altogether. These historical explorations also encouraged P. C. Roy and J. C. Bose to believe that non-human life forms are capable of experiencing pleasure and pain.

We considered four of these scientists in more detail. P. C. Roy attributed his chemical discoveries to intuition or *anubhava*, a term used in *advaita* Vedānta, but also familiar to members of the Brahmo Samaj. J. C. Bose demonstrated the polarization of electromagnetic waves and researched parallels between the behaviour of metal coherers and plant and animal responses to external stimuli. He set up the Bose Institute in 1917 dedicated to exploring the frontiers between the different sciences, thereby uncovering the 'One' underlying natural phenomena, and was praised by Rabindranath Tagore for delineating the linkages between the living and the non-living.

S. N. Bose undertook outstanding work on both particle physics and field theory, collaborating extensively with Einstein on general relativity and superconductivity. Like Einstein he was motivated by the search for a unified field theory, a quest which he associated with his belief in *advaita* Vedānta. Srinivasa Ramanujan was a brilliant mathematician who never ventured into other areas of research, but whose collaborative efforts with G. H. Hardy eventually bore extensive fruit in the physical sciences. His work was partially motivated by his belief in *advaita* Vedānta.

All these scientists interpreted their work as the expression of their religious and philosophical beliefs, as did several others that we have briefly considered, the only major exception being Meghnad Saha, who reacted against the discriminatory treatment he received from caste Hindus, and felt that the Hindu scriptures contained too many inconsistencies to be taken seriously.

7 An investigation into the beliefs of Indian scientists

A group of postgraduate scientists at the Indian Institute of Science in Bangalore are discussing the relationship between religion and science. They are a Muslim from Hyderabad (M), two Saraswat brahmins (Sb1 and Sb2), a Sikh (S), a brahmin from North India (B) and a woman from Tamil Nadu, who is one of the relatively small number of non-brahmins at the Institute (Nb) – she did not wish to be more specific about her sub-community. I shall refer to myself as DG.

M: You can broadly speak of two schools of thought, the one concerned with mind and the other concerned with matter. The first admits the supernatural and revealed knowledge. Science prefers the second. The scientist believes in revealing knowledge.

Sb1: For me the issues are less academic. My parents were very orthodox when I was young. But the strength of religion is decreasing. My children will not practise religion at all. Therefore I am in an intermediate position.

Nb: Many of us in this Institute are double people. As scientists we are rational, but when we leave the laboratory and go home we behave differently.

M: Science has certainly made very little impact on the home and social life in general. Perhaps we tend to think of it as western.

Sb2: How can science be eastern or western?

Nb: It can't – but the religion of a particular society can conflict with science, and there are differences between East and West. Galileo experienced conflict with the Church: Aristotle tried to maintain his religious ideas, and they hampered his science.

DG: It's interesting to hear Galileo and Aristotle being mentioned. Do you feel caught up in the conflicts between science and religion in the West?

M: It is not for me just a question of social and cultural comparisons. As a Muslim I believe that God is almighty and Muhammad is his prophet. If I cannot reconcile either part of this statement with reason I will reject it. Then I feel no conflict.

S: But religion is not only doctrinal. Where does moral conduct come in?

Nb: Morality and God are different questions altogether. All religions aim to avoid sin and do good. Social behaviour among the higher castes has been influenced more than anything else by western thought. Inter-caste

Beliefs of Indian scientists 103

marriages now take place frequently and that is a good thing which has come from western influence.
Sb1: Can I suggest that we discuss our own campus?
Sb2: What are the indicators of religious beliefs in the Institute and to what extent are they harmful or beneficial to us as scientists?
Sb1: The *pūjā* holidays – especially *āyudha pūjā*; Sarasvatī *pūjā*; visits to churches and temples just before exams; Dīwālī fireworks.
Nb: Crackers at Dīwālī – does anyone object?
M: It's just for fun!
B: There are very strong social pressures on anyone who wishes to work during the *pūjā* holidays. This year the Registrar sent a circular asking that *āyudha pūjā* should only be performed after four in the afternoon.
Nb: Once a year at least the machines in the Institute get cleaned properly!
M: As a Muslim why should I observe *pūjā*? If a Hindu wanted to he could work in the laboratory, say, during Christmas, but it is impossible for anyone to work during Hindu *pūjās*. One head of department insists that if you do not wish to take part in the *pūjā* you must give a written apology in advance. For a Christian or Muslim this is intolerable!
Sb2: One person should not interfere with another with regard to rites. I have been in both Christian and Muslim institutions and in neither was I forced to practise their religion.
Nb: I went to a Convent school and pressure was put on us there!
Sb: It seems to me that we must separate the religious and social factors before anything else.

The investigation into the beliefs of Indian scientists was conducted at four centres using questionnaires and interviews. The questionnaire, which was preceded by a trial questionnaire, is included as Appendix B. It was distributed and collected by senior faculty members at the scientific institutions. The interviews, extracts from which will be quoted, were conducted by the author. The results are available in a research series published in Bangalore and Chennai.[1]

The universities and science institutes at which the investigation was conducted were located in Delhi (where I had previously taught physics at the university for four years), Bangalore (where I was based at the Indian Institute of Science (IISc) for the two-year duration of the investigation), Madurai (Tamil Nadu), which is more regionally South Indian than the Bangalore IISc, and Kottayam (Kerala), where there is a high proportion of Christians.

Framing the hypotheses

Although the primary purpose of the investigation was to study the impact of science upon religion and the extent to which respondents related their religious and scientific beliefs, an attempt was also made to determine attitudes to issues which had played an important role historically – particularly evolution, cosmology and scientific–historical approaches to scripture. At an early stage of the investigation

104   Beliefs of Indian scientists

it was felt that the doctrine of reincarnation, though never seriously questioned in the nineteenth century (except by Ram Mohan Roy), was increasingly being challenged as a result of exposure to scientific ideas. It was therefore included in the questionnaire and discussed during the interviews. The crucial issues were whether or not science students and research scientists consciously related religion and science, the nature of the relationship, if any, and the manner in which exposure to scientific ideas influenced their religious beliefs.

It was felt that the proposed questionnaire and interview investigation would be more objective if it could be designed so as to test specific hypotheses dealing with the relationship between science and religion, and after a trial questionnaire had been distributed at the IISc, the following hypotheses were framed:

Hypothesis (a)   Science has a negative effect upon the strength of religious beliefs among scientists.

Hypothesis (b)   The greater the amount of scientific study the greater the rate of rejection of religion by scientists.

Hypothesis (c)   The degree of perceived conflict between science and religion has an inverse correlation with the importance attached to religion. Thus a high degree of perceived conflict is related to a low valuation of religion, and a low degree of perceived conflict is related to a high valuation of religion.

The first two hypotheses were framed on the basis of preliminary discussions with research scientists at the IISc and the generally held view of many sociologists that whenever science and religion come into any kind of encounter, science will eventually predominate. In order to measure the 'perceived conflict' between science and religion as required to test the third hypothesis, a number of questions were framed, some of which referred specifically to issues which had played a significant role historically. Thus 'perceived conflict' was measured by responses to the following:

3. (a) Do you think that there is any conflict between your own religion and science?   Yes / No
   (b) Has your degree course of study changed or modified your religious beliefs at all?   Yes / No   In what way?
4. Which, if any, of the following *do not* agree with a religious outlook?
   (a) Biological evolution   (d) The use of reason
   (b) Theories of universe's origin (e) The necessity for proof
   (c) Technological progress   (f) Other............
5. Which, if any, of the following *do not* agree with a scientific outlook?
   (a) Existence of the soul   (d) Reincarnation
   (b) Prayer   (e) Miracles
   (c) Life after death   (f) Other............
6. Indicate with a short statement how you think science and religion are related.[2]

The responses to these questions were interpreted so as to obtain an estimate of the degree of perceived conflict between science and religion.

The importance attached to religion was measured from responses to the following:

10. Religion is important to me..........
    (a) At all times
    (b) Family occasions
        (e.g. weddings, funerals)
    (c) When I need help
    (d) Never

It was expected that respondents in the first category (religion important 'at all times') would give significantly different answers to the rest of the questionnaire from those in the last category (religion 'never' important).

In order to establish that science was associated with a change of religious beliefs, responses to question 3(b) were coded according to whether the study of science had increased or diminished religious belief. The possibility that changes in outlook attributed to science might equally well be explained by other factors such as age was also investigated, and additional relevant information was therefore requested in the first part of the questionnaire. Details about attendance at places of worship were sought in order to see whether or not a correlation existed between the importance attached to religion, which involves a subjective judgement, and an external, more objective criterion. While the absence of such a correlation would not necessarily invalidate the hypotheses since many devout Hindus do not attach much importance to regular visits to places of worship, the existence of a correlation would serve to strengthen the hypotheses.

Previous research

Apart from general references to assumed inconsistencies between scientific and religious beliefs and attitudes, the most careful sociological investigations are contained in *Science, Technology and Culture*, published by the Indian Research Council for Cultural Studies.[3] From the point of view of formulating the basic problem and scope of the investigation which was undertaken, the most important sections of this book are a study of social constraints on science in India by Purnima Sinha, a chapter on the social characteristics and attitudes of a group of Sri Lankan research scientists by D. L. Jayasuriya, and the conclusions of a number of interviews with Indian scientists by Surajit Sinha.

Purnima Sinha's observations are based largely on her research and experience in a number of physics laboratories in India and the USA. She considers the hypothesis of Brijen Gupta, professor of history at New York University, that the 'other-worldly attitude' of India's religions has inhibited and continues to inhibit attitudes to scientific research and development. Her main contention is that while it is extremely difficult to establish any meaningful relationship between the Hindu tradition and the observable constraints on scientific research in Indian laboratories, many concrete non-religious factors which play a negative role in the development of science may be identified. Surajit Sinha's conclusions are based upon interviews with sixty-seven scientists, most of them Hindus aged between 30 and 40. He refers to the preponderance of brahmins in many scientific institutions, and describes the economic background of his sample as middle-class. Several of his interviewees were conscious of a gap between their scientific and religious worldviews in the laboratory and at home.

The social backgrounds of Indian scientists have been the subject of a survey by Ashok Parthasarathi, who concludes that more than 70 per cent of all Indian scientists come from homes where the father is engaged in a non-scientific profession.[4] It is therefore not surprising to find a certain amount of conflict between the values encountered by many scientists at their places of work and in their homes.

Several publications in this area by the author are listed separately.[5] There are also a number of unpublished ones which were destined for *Contributions to Indian Sociology*, edited by Louis Dumont and David Pocock. Unfortunately, this excellent journal ceased publication for a number of years (probably on account of the illness of Professor Dumont), though it is currently being published again under the editorship of Professor Patricia Uberoi of the Institute of Economic Growth in Delhi.[6]

Results of the investigation

Five major institutions were visited, the Indian Institute of Science (IISc), Bangalore, Delhi University, the American College, Madurai and the Church Missionary Society (CMS) College, Kottayam. Of these institutions the first was selected on account of its high standards and the regional spread of its members. The three colleges of Delhi University, two men's, one women's, and the Delhi IIT were chosen because they contained representatives of a wide variety of largely urban and northern graduates and undergraduates. In comparison with the IISc where brahmins tended to be in the majority, the American College, Madurai, was known on account of its admission quota procedures to contain a high proportion of non-brahmins. This situation is likely to change drastically following the government's decision in April 2006 to implement the recommendations of the Mandal Commission relating to reservations for members of Other Backward Classes.[7] The CMS College in Kottayam was selected in the expectation that there would be a high proportion of Christians. The American and CMS colleges were also felt to be characteristic of a less urban and cosmopolitan type of membership than the others. Approximately 800 questionnaires were distributed by faculty members at the institutions listed in Table 7.1. In most cases respondents were permitted to complete and return them in class. Postgraduates were contacted via their heads of department. Just under 700 questionnaires were completed, and 155 interviews were conducted by the author. Questionnaire responses were coded and investigated statistically using the IBM 360 computer at the IISc.

Details of the social, educational and regional background of the respondents were obtained from the first part of the questionnaire, and are summarized elsewhere.[8] The Bangalore IISc respondents tended to be in their early to mid-twenties, whereas those at the other centres were younger (the samples also included junior faculty members). The majority of the Delhi respondents (77 per cent) had been to school in

Table 7.1 Courses and institutions of the respondents

College or institution	Class	Code number
Kirioli Mal College, Delhi University	First year BSc General	1-1001–1-1021
	Second year Chemistry Honours	1-1022–1-1041
	Pre-medical	1-1042–1-1087
	Third year BSc General	1-1088–1-1117
St Stephen's College, Delhi University	First year Chemistry Honours	1-2118–1-2141
	First year BSc General and Honours Maths	1-2142–1-2158
	Third year BSc General	1-2159–1-2187
	Non-science (mostly Honours English)	1-2188–1-2200
Delhi University non-collegiate postgraduates	MSc and PhD	1-3201–1-3224
Miranda House, Delhi University	Third year BSc General	1-4225–1-4269
Indian Institute of Technology, Delhi	BTech (first, second, fourth and fifth years)	1-270–1-5382
Indian Institute of Science, Bangalore	BEng, MEng, and PhD	2-384–2-582
American College, Madurai	First year MSc Zoology	3-1583–3-1594
	Second year MSc Physics	3-2595–3-1607
	Second year MSc Chemistry	3-1608–3-1621
	First year MSc Chemistry	3-1622–3-1636
Church Missionary Society College, Kottayam	Second year MSc Chemistry	3-2637–3-2648
	First year MSc Botany	3-2649–3-2658
	Third year BSc Chemistry	3-2659–3-2690

Delhi, and the same proportion stated that their parents lived in a city – usually Delhi, though in some cases Kolkata (formerly Calcutta).

Whereas only one per cent of the Delhi respondents came from the south, 23 per cent of the Bangalore IISc sample specified a native place in North India – this is consistent with the Institute's policy of giving scholarships and accommodation on an all-India basis. The remaining IISc respondents came from Mysore (29 per cent), Tamil Nadu (22 per cent), Kerala (13 per cent) and Andhra Pradesh (13 per cent). Most of the Kottayam and Madurai respondents came from their local state. In agreement with Ashok Parthasarathi's findings most of the fathers at all four centres were in non-scientific occupations.[9] The fact that 58 per cent of the fathers of Kottayam respondents were in non-scientific jobs in the private sector (as compared to 15–35 per cent at the other three centres) is because the majority were Christians, who in Kerala control a sizeable share of private sector undertakings.

For historic reasons Christians own and administer a large number of schools, colleges and hospitals throughout India, admission to which is much sought after because of their high standards.[10] Thus the CMS College in Kottayam – usually referred to only by its acronym – was founded by the British Church Missionary

108 *Beliefs of Indian scientists*

Society (now renamed the Church *Mission* Society on account of the colonial overtones of the word 'missionary'). St Stephen's College in Delhi also maintains consistently high academic standards, as do many of the schools run by Roman Catholics. We have therefore classified the responses according to whether the high schools attended were Christian, religious non-Christian (e.g. run by the Ramakrishna Mission) or non-religious (i.e. state). The breakdown is shown in Table 7.2.

Religious affiliation

Respondents were asked to specify both their personal religion and that of their parents. The results are shown in Table 7.3. Most of the Delhi respondents classified under the heading 'other' were Sikhs, the remaining few in Delhi and elsewhere being Buddhists, Jains, Parsis or members of sects.

A comparison between personal and parents' religion gives an indication of the extent to which representatives of different religious traditions have ceased to identify themselves with their original community. Muslims in Bangalore showed a particularly marked trend in this respect, though elsewhere the 'no religion' respondents were drawn mainly from Hindu and, to a lesser extent, Christian backgrounds. It was clear from the interviews that the category 'no religion' was less emphatic than the response 'religion is never important' in part B of the questionnaire. Some respondents simply felt that the highest form of belief transcends identification with any particular religious group, and for this reason stated that they had no (particular) religion. A few specified 'humanism' under the category 'other'. A very small percentage (1.0 per cent) of Delhi respondents stated that

Table 7.2 Religious affinity of high school attended (percentages)

	Religious Christian	*Religious non-Christian*	*Non-religious*
Delhi	26.4	6.6	67.0
Bangalore	20.4	9.5	70.1
Kottayam	60.0	8.0	32.0
Madurai	44.6	10.8	44.6

Table 7.3 A comparison between personal and parents' religious affiliation (percentages)

		Hindu	*Muslim*	*Christian*	*Other*	*No religion*
Delhi	Personal	71.1	0.9	3.3	10.9	13.8
	Parents	82.0	1.0	4.0	12.0	1.0
Bangalore	Personal	75.4	2.5	4.0	3.5	14.6
	Parents	87.0	4.7	4.7	3.6	—
Kottayam	Personal	27.8	1.9	57.4	—	12.9
	Parents	37.0	1.9	61.1	—	—
Madurai	Personal	62.9	3.7	27.8	—	5.6
	Parents	68.5	3.7	27.8	—	—

neither they nor their parents belong to any religion. Elsewhere parents' religion was consistently specified under one of the first four headings.

Brahmins at the IISc were identified from responses such as Smārtha, Iyer and Nambūdri to the question requesting them to specify their religious sub-group. Similarly, non-brahmins in Madurai could sometimes be identified from their sub-caste and family names. It was estimated from the interviews and other sources of information that between 75 and 90 per cent of all students at the IISc were brahmins.

The relationship between science and religion

In this section we consider the overall responses to the second part of the questionnaire, illuminating these where appropriate with comments made in the interviews. Interviewees may be identified by referring their numbers to Table 7.1. The statistical analysis of the data will be discussed later; percentages expressed to three significant figures were generally found to be valid from the statistical tests.

Responses to the question 'Do you think that there is any conflict between your own religion and science? Yes/No' (question 3a) are summarized in Table 7.4. The distributions for all four groups are fairly similar. Table 7.5 shows the responses to the second part of the same question: 'Has your degree course of study changed or modified your religious beliefs at all? Yes/No (question 3b). In what way?' The distribution of responses is fairly similar to the previous table except that a much larger proportion of Bangalore respondents (57 per cent) stated that their studies had changed their beliefs. It was clear from the interviews that this is because they had spent longer studying and researching in science and were mostly committed to a scientific career.

Table 7.4 Whether conflict perceived between religion and science (percentages)

	Yes	No
Delhi	26.6	73.4
Bangalore	30.0	70.0
Kottayam	30.0	70.0
Madurai	35.6	64.4

Table 7.5 Influence of degree course (percentages)

	Yes	No
Delhi	28.7	71.3
Bangalore	57.0	43.0
Kottayam	42.1	57.9
Madurai	35.2	64.8

110 *Beliefs of Indian scientists*

The direction of the changes in religious belief as indicated by responses to the open-ended part of question 3 is shown in Table 7.6. Further clarification of responses to the open-ended part of this question was sought in the interviews. 'Beliefs discarded or reduced' and 'beliefs initiated or increased' may be illustrated by the following two statements:

> I have rejected religion since doing pre-medical studies.... It is wrong to think that good parents will have children and bad parents will be denied children by God. It just isn't true. Genetics determines what sort of children you have and not God. (1-4228)
>
> Science has increased my beliefs and given me a sense of wonder. (2-396)

The category 'beliefs modified' implied that a respondent saw the need to relate science and religion, and often included an element of doubt, so that on the whole the effect of science upon religious belief was negative. However, the presence of a significant percentage who stated that the study of science had increased their beliefs was not consistent with the first of the three initial hypotheses, that is, science has a negative effect upon the strength of religious beliefs among scientists.

For convenience the responses to questions 4 and 5 in part B of the questionnaire will be discussed together. The percentages of responses to different parts of these two questions at each of the four centres visited are shown in Table 7.7. The figure (1) at the top of a column indicates no conflict and (2) indicates conflict. Thus, for example, 54 per cent of the Delhi sample felt that biological evolution does not conflict with a religious outlook, whereas 46 per cent felt that it does.

Within the four groups there was a steady variation in the responses to different parts of the two questions. Technological progress and prayer were consistently the cause of less conflict than reincarnation and life after death. The category 'other' was seldom marked, and miracles, the use of reason and proof produced a fairly steady distribution between 'conflict' and 'no conflict'. Biological evolution and theories of the universe's origin were the cause of conflict for a sizeable proportion of both Kottayam and Madurai respondents. This may be due to the higher proportion of Christians who have problems reconciling Darwinism and their religious belief.

Table 7.6 Direction of change in beliefs due to influence of degree course or scientific research (percentages)

	Beliefs discarded or reduced	Beliefs modified	No change	Beliefs initiated or increased
Delhi	12.4	5.2	78.4	4.0
Bangalore	28.2	12.8	43.6	15.4
Kottayam	13.7	13.7	58.9	13.7
Madurai	5.5	18.5	64.8	11.2

Table 7.7 A comparison of responses to specific areas of possible conflict between science and religion (percentages)

	Delhi		Bangalore		Kottayam		Madurai	
	(1)	(2)	(1)	(2)	(1)	(2)	(1)	(2)
Science								
Evolution	54.0	46.0	60.0	40.0	38.9	61.1	39.6	60.4
Universe's origin	63.0	37.0	62.0	38.0	38.9	61.1	49.0	51.0
Technological progress	69.8	30.2	79.0	21.0	90.7	9.3	86.9	13.1
Reason	73.5	26.5	78.5	21.5	70.4	29.6	77.1	22.9
Proof	58.0	42.0	66.1	33.9	57.4	42.6	56.7	43.3
Other	97.0	3.0	94.5	5.5	98.2	1.8	96.1	3.9
Religion								
Existence of soul	48.1	51.9	53.8	46.2	48.1	51.9	54.8	45.2
Prayer	61.7	38.3	72.0	28.0	74.1	25.9	71.8	28.2
Life after death	38.5	61.5	47.1	52.9	35.2	64.8	43.5	56.5
Reincarnation	59.0	41.0	48.5	51.5	40.7	59.3	54.8	45.2
Miracles	53.9	46.1	48.8	51.3	53.7	46.3	58.6	41.4
Other	93.1	6.9	91.5	8.5	98.2	1.8	94.0	6.0

Note
(1) means 'no conflict', (2) means 'conflict'.

Table 7.8 Importance attached to religion (percentages)

Religion important	Delhi	Bangalore	Kottayam	Madurai
At all times	48.0	56.5	74.1	72.2
Family occasions	21.0	16.8	5.5	11.2
When I need help	22.3	11.0	3.7	9.3
Never	15.8	21.0	18.5	11.2

The responses to several of the questions which have been discussed so far were intended to measure the degree of perceived conflict between science and religion. The other variable in the third of the three initial hypotheses is the importance attached to religion, and this was measured by the first part of question 10 in part B of the questionnaire. It was anticipated that those for whom religion was 'never' important would give significantly different answers to the earlier questions than those for whom religion was important 'at all times'. The percentages of those giving a positive indication to the four alternatives are shown in Table 7.8. Delhi respondents showed the greatest inclination to specify the intermediate responses 'family occasions' and 'when I need help', and a few respondents opted for both these categories. The proportion of those for whom religion was 'never' important was greatest for Bangalore and least for the Madurai respondents.

The remaining questions in part B of the questionnaire are not directly relevant to the initial hypotheses. A few of them will be discussed briefly. The responses to

112 *Beliefs of Indian scientists*

Table 7.9 Extent of agreement with Einstein's statement (percentages)

| | Agree with | | | |
	Both parts	First part	Second part	Neither
Delhi	35.0	12.1	22.9	30.0
Bangalore	49.0	9.2	18.8	23.0
Kottayam	52.0	14.0	20.0	14.0
Madurai	46.2	21.1	11.6	21.1

Table 7.10 The importance of different authorities in making ethical decisions (percentages)

	Delhi	*Bangalore*	*Kottayam*	*Madurai*
Philosopher	17.6	19.1	24.5	8.5
Scientist	40.6	32.1	30.5	32.0
Magazine articles	16.8	10.7	16.4	8.5
Religious leader	7.5	7.1	14.2	36.1
Politician	1.4	1.8	6.2	4.3
Other	16.1	29.2	8.2	10.6

Einstein's statement in question 2, 'Science without religion is lame; religion without science is blind', are shown in Table 7.9. Madurai respondents were inclined to feel that science needs to be supplemented by religion; elsewhere the tendency was to feel that religion needed to be supplemented by science.

Question 7 was designed to try to find out what sort of authority respondents might turn to in order to make a moral or ethical choice:

> 7. If you needed help on a *moral or ethical* issue related in some way to scientific progress (e.g. birth control, the use of nuclear weapons), to which *one* of the following sources would you be most likely to turn?
> (a) Philosopher (d) Religious leader
> (b) Scientist (e) Politician
> (c) Magazine articles (f) Other..........

The responses are shown in Table 7.10. Many Bangalore scientists who opted for 'other' specified that they would think things out for themselves – a response which is not surprising from a slightly older group. From the interviews it appeared that there was a certain amount of overlap between some of the alternatives. Thus, for example, Bertrand Russell was given as an example of both a philosopher and a scientist, though generally speaking a philosopher was understood to be a person who thought rationally but whose training went beyond the academic study of science. The category 'magazine articles' was investigated further in the interviews and included a wide range of literature. *Bhavan's Journal*, the weekend supplements of the English-medium national dailies (e.g. the *Indian Express*, the *Times of India* and the *Hindu*) were all extremely popular. In the north *Sarita* was quite extensively read, and among Kerala respondents the *Malayala Manorama* was influential. *Blitz*, a Bombay-based *News of the World* weekly tabloid, was read at all four centres visited, though few respondents were inclined to take it too seriously.

Examples of religious leaders given by those who specified part (d) of question 7 varied considerably, and ranged from the village priest or family *guru* to well-known contemporary or historical figures. Swami Ranganathananda was sometimes given as a modern example, and Vivekananda and Ramakrishna were cited as outstanding religious leaders of the past. Christians tended to identify religious leaders with their local priest or minister, and where magazine articles were specified by them the examples given often included literature produced by the Evangelical Union and similar organizations. With specific reference to birth control some Muslims stated that the choice between 'religious leader' and 'scientist' was a difficult one.

The lack of concern for ritualistic worship is reflected in the response to question 11. The responses to the second part of the question are given in Table 7.11. The large percentage of Kottayam respondents who worshipped weekly (72 per cent) is consistent with the fact that approximately 60 per cent of them were Christians. About 30 per cent of the Madurai sample were also Christians, and if these two percentages are each subtracted from the respective totals of those who worshipped weekly (72 and 57.5 per cent), the resulting percentages are of the same order as those of the Bangalore and Delhi groups which were composed mostly of Hindus. A comparatively high percentage of Madurai respondents worshipped daily. From the interviews it was clear that daily worship often meant prayer, meditation or *sūryanamaskāram* (i.e. acknowledging daybreak). From the interviews it was also concluded that the high proportion of Delhi and Bangalore respondents who marked 'special occasions only' were referring to family ceremonies, times of crisis and anxiety (e.g. before exams), and, in the north, the *pūjā*s.

The responses to the second part of question 10, 'explain the meaning of God to you', are shown in Table 7.12. The distinctions between 'personal',

Table 7.11 Frequency of attendance at places of worship (percentages)

	Daily	Weekly	Fortnightly	Monthly	Special occasions	Annually
Delhi	10.0	19.2	3.9	7.0	56.0	3.9
Bangalore	11.0	29.0	6.0	10.0	42.0	2.0
Kottayam	9.3	72.0	4.7	4.7	7.0	2.3
Madurai	17.5	57.5	10.0	10.0	5.0	—

Table 7.12 Interpretation of God (percentages)

	Personal	Not personal	Unspecified	Don't believe
Delhi	30.7	26.1	27.6	15.6
Bangalore	26.1	22.6	33.7	17.6
Kottayam	42.0	20.0	30.0	8.0
Madurai	60.3	18.9	15.1	5.7

'not personal' and 'unspecified' were difficult to apply consistently, and possibly the only significant conclusion which may be drawn from this table is that fewer Kottayam and Madurai respondents put themselves in the 'don't believe' category than in Delhi and Bangalore. Some interesting and original definitions of God were given, as may be seen from the following examples:

> God is not Krishna or Muhammad or Christ or Buddha to me. God is someone supernatural who has created us and sent us onto this good earth saying 'Come on, I want you to do something for your fellow beings and for me.' (Hindu, 2-412)
>
> God is one for whom one cannot put models. God is animating all living beings and underlying everything that happens. (Hindu, 3-1631)
>
> God is omnipotent and creates, destroys and saves all lives. (Hindu, 3-1611)
>
> God is a boundary within which you remain human. (Hindu, 2-469)
>
> God is absolute power, knowledge and eternal bliss. (Hindu, 2-474)
>
> God is my only Master. (Muslim, 3-1624)
>
> I believe God is my Creator. But for belief in God life would have been meaningless for me. The belief in the Creator compels me to learn more and do something for others, his creatures and my neighbours. (Christian, 3-2639)
>
> God is the creator and destroyer of the world. (Christian, 2-562)

These definitions are not particularly representative of the religious tradition to which the respondent belonged.

Investigation of the hypotheses

It is clear from the general data presented so far that the first hypothesis, namely, that science has a negative effect upon religious belief, is not correct. In some cases it enhances it, and from the interviews it is also clear that the consequent increase in religious belief does not for the most part consist of reassertive fundamentalism. A significant number of respondents from all the major religious communities described their increase of belief in adaptive and, occasionally, highly original terms. The first hypothesis was therefore reformulated.

The second hypothesis, implying a direct correlation between the amount of scientific study and the rate of rejection of religion by scientists, was difficult to investigate because the length of scientific training at university level could not always be determined. Also it appeared from analysis of the third hypothesis that age was not a particularly significant factor in the degree of importance attached to religion, and since the amount of scientific study might be expected to be a function of age, this tended to make the second hypothesis less credible. Graphs showing the importance attached to religion as a function of age are available elsewhere, and it is clear from inspection that there is little variation in response.[11] No statistical analysis was therefore made of the first two hypotheses.

Beliefs of Indian scientists 115

According to the third hypothesis a high valuation of religion should be related to a low degree of conflict between science and religion, and a low valuation should be related to a high degree of conflict. The valuation of religion was measured by the comparative responses to question 10 of part B of the questionnaire, particularly the first and fourth alternatives, that is, religion is important 'at all times' and 'never'. It was expected that the intermediate alternatives ('family occasions' and 'when I need help') would form some sort of continuum within these two extremes. Degrees of conflict were measured by questions 3, 4, 5 and to some extent 6.

It is not possible to cover all the relevant data, but a general indication of the different pattern of responses according to whether respondents considered religion important 'at all times' or 'never' may be seen from the following example. In question 5, reincarnation was specified as a possible area of conflict between religion and a scientific outlook. The distribution of responses for the Bangalore sample to this part of the question in relation to the importance attached to religion is shown in Table 7.13.

It is clear from this table that the first three alternative responses were distributed fairly consistently between 'no conflict' and 'conflict'. The fourth alternative, 'never', gave a very different distribution suggesting that respondents in this category saw much more conflict between reincarnation and science than the others. The probability that such a distribution could have occurred by chance may be obtained by calculating Chi Square, which in this case is 12.70. The odds are therefore better than one in a hundred that this distribution could have occurred by chance, and when the first three alternatives – 'at all times', 'family occasions', and 'when I need help' – are combined, the level of statistical significance becomes better than one part in a thousand.[12]

This particular example is given to illustrate the manner in which different parts of the questionnaire have been used to establish a relationship between the importance attached to religion as estimated by the responses to question 10, and the degree of perceived conflict as measured by the responses to questions 3, 4 and 5. Detailed calculations of Chi Square were carried out with the IBM 360 computer at the Indian Institute of Science. In each case the computer was programmed to give Chi Square for both three degrees and one degree of freedom, that is, first with the four alternative responses to question 10 considered separately, and then with the first three telescoped ('at all times', 'family occasions', and 'when I need help'), and the fourth (i.e. 'never') on its own.

Table 7.13 Distribution of responses to reincarnation as a function of the importance attached to religion (percentages)

Religion important	No conflict between reincarnation and science	Conflict between reincarnation and science
At all times	53	47
Family occasions	58	42
When I need help	58	42
Never	23	77

116 *Beliefs of Indian scientists*

The value of Chi Square and the corresponding levels of statistical significance for three degrees and one degree of freedom for the two parts of question 3 at all four centres are shown in Table 7.14. For three degrees of freedom (df = 3), Chi Square varies from 1.86 to 11.17, giving a corresponding range in the significance level from 0.7 to beyond 0.02. By convention levels greater than 0.5 (the second and fourth columns) are considered to be statistically insignificant, so that the first and last values of the first column are inconclusive evidence in support of the third hypothesis. (Chi Square should be high; the corresponding level of statistical significance should be low.) For one degree of freedom (df = 1) Chi Square fluctuates from zero to 5.79, the latter corresponding to a statistical level of better than 0.02. The low values of Chi Square for Madurai and Kottayam responses in this table and in subsequent ones are due to poor statistics for all categories except religion important 'at all times'.

The values of Chi Square are much higher for questions 4 and 5. Selecting 4a (biological evolution) and 5d (reincarnation) as examples, Chi Square and the corresponding levels of significance are as shown in Table 7.15. Chi Square is

Table 7.14 Chi Square and the levels of significance for perceived conflict between religion and science and influence of degree course

		Three degrees of freedom (df = 3)		One degree of freedom (df = 1)	
		Chi Square	Level of significance	Chi Square	Level of significance
Delhi	3(a)	1.86	0.7	1.26	0.3
	(b)	5.33	0.2	1.72	0.2
Bangalore	3(a)	8.22	0.05	5.79	0.02
	(b)	5.40	0.2	2.01	0.2
Kottayam	3(a)	5.46	0.2	2.50	0.2
	(b)	9.80	0.05	1.38	0.3
Madurai	3(a)	11.17	0.02	5.61	0.02
	(b)	2.32	0.7	0.00	—

Table 7.15 Chi Square and the corresponding levels of significance for evolution and reincarnation

		Three degrees of freedom		One degree of freedom	
		Chi Square	Level of significance	Chi Square	Level of significance
Delhi	Evolution	13.30	0.01	2.40	0.2
	Reincarnation	4.70	0.2	1.23	0.3
Bangalore	Evolution	16.37	0.001	12.87	0.001
	Reincarnation	12.70	0.01	12.37	0.001
Kottayam	Evolution	5.10	0.2	0.51	0.5
	Reincarnation	0.17	0.98	0.02	0.9
Madurai	Evolution	6.47	0.1	1.75	0.2
	Reincarnation	12.53	0.01	3.51	0.1

low for the Kottayam and Madurai groups, and reducing the number of degrees of freedom does not improve the level of significance. Chi Square for reincarnation for Delhi respondents is surprisingly low, though significant (better than 0.2). Provided the four alternatives to question 10 may be assumed to give a reliable indication of the importance attached to religion, the third hypothesis, namely, that the degree of perceived conflict between science and religion has an inverse correlation with the importance attached to religion, would appear to be valid.

While the importance attached to religion need not necessarily correlate with attendance at places of worship (especially for Hindus), the four alternatives – 'at all times', 'family occasions', 'when I need help', and 'never' – were analysed to see how they correlated with question 11 relating to frequency of attendance at places of worship. The results are shown in Table 7.16.

Those for whom religion is important 'at all times' tend to visit places of worship, whereas those for whom religion is 'never' important for the most part do not. The category 'special occasions only' for the second part of the question was sometimes associated with an affirmative and sometimes with a negative response to the first part. Further breakdown of the responses showed that the two alternatives 'family occasions' and 'when I need help' are characteristic of respondents who visit places of worship much less frequently than those in the 'at all times' category. This provides an additional reason for retaining the four original groups and performing the statistical analysis for three degrees rather than for one degree of freedom.

To sum up – the first two initial hypotheses were found to be in need of reformulation and hence the data was not subjected to any statistical tests in relation to them. The third hypothesis, stating that an inverse correlation exists between the degree of perceived conflict between science and religion and the importance attached to religion, was shown to be valid, though more so for some areas of perceived conflict than for others. It was also demonstrated that the importance attached to religion can be related to the frequency of attendance at places of worship and that those for whom religion was only important on family occasions or at times of crisis had a different pattern of attendance at places of worship from those for whom religion was important at all times. This conclusion strengthened

Table 7.16 Importance of religion analysed in terms of attendance at places of worship (percentages)

Religion important	Attendance at place of worship	
	Yes	No
At all times (357 responses, 51.6% of total)	84.2	15.8
Family occasions (116 responses, 16.8% of total)	76.1	23.9
When I need help (107 responses, 15.4% of total)	73.9	26.1
Never (112 responses, 16.2% of total)	29.1	70.9

the case for analysing the first three categories of question 10 separately, and the fact that Chi Square did not increase appreciably for one degree of freedom was consistent with this. It can also be demonstrated that age did not play a significant role with regard to the importance attached to religion.

In the next section it will be shown that the third hypothesis remains valid when the entire sample from all the four centres visited is sub-divided into the two major religious sub-groups and the same statistical tests are applied.

Comparison between Hindus and Christians

When the responses at all four centres were analysed according to religion of parents (question 19, part A), the distribution was as follows: 526 Hindus, 15 Muslims, 70 Christians, 45 'other', 4 no religion.

The Hindu and Christian responses were analysed in separate groups to see whether or not the hypotheses still held good. The majority of both groups stated that they attended places of worship, the actual proportion being 98 per cent for Christians and 82 per cent for Hindus. Eighty-seven per cent of Christian respondents attended places of worship weekly, whereas the Hindu frequency of attendance was much more widely distributed. These figures imply that the significance of the response that religion is important 'at all times' is not exactly the same for Hindus and Christians.

The pattern of responses to the questions pertaining to conflict between science and religion was very similar for the two groups. This was particularly apparent in relation to question 3:

3. (a) Do you think that there is any conflict between your own religion and science?   Yes/No
   (b) Has your degree course of study changed or modified your religious beliefs at all?   Yes/No   In what way?

Seventy-eight per cent of Hindus in the 'at all times' category responded negatively to the first part, and 70 per cent replied similarly to the second part, whereas for Christians the corresponding figures were 74 per cent and 68 per cent respectively.

A greater proportion of Hindus in the 'at all times' category consistently saw no conflict between science and religion with regard to biological evolution, theories of the universe's origin, technological progress, the use of reason, the necessity for proof, the soul, prayer, reincarnation and miracles (questions 4 and 5). The only exception to this trend was life after death (question 5c) for which the majority of Hindus (54 per cent) felt that there is conflict. From the interviews it was clear that they did not make a connection between reincarnation and life after death.

A greater proportion of Christians for whom religion was important 'at all times' consistently saw no conflict between science and religion with regard to theories of the universe's origin, technological progress, the use of reason, the necessity for proof, the soul, prayer and miracles. The three exceptions were biological evolution, life after death and reincarnation where the majority

Table 7.17 Values of Chi Square for Hindus and Christians according to the degree of importance attached to religion for a selection of responses to questions 4 and 5

	Hindus		Christians	
	$df = 3$	$df = 1$	$df = 3$	$df = 1$
Evolution	20.29	11.91	7.59	1.12
Universe's origin	29.88	19.32	6.40	1.08
Life after death	18.01	14.00	2.65	0.50
Reincarnation	11.03	10.90	1.83	0.04

(62, 53 and 53 per cent respectively) saw a conflict in these areas. As has already been pointed out, although Hindus tended to differentiate between reincarnation and life after death, Christians did not appear to do so. For biological evolution, theories of the universe's origin, life after death, and reincarnation, the values of Chi Square for three degrees and one degree of freedom (df = 3, df = 1) are shown in Table 7.17.

For Hindus Chi Square is never less than 11.03, corresponding to a level of significance better than 0.02 for three degrees of freedom. The statistical level is improved for one degree of freedom (better than 0.001). For Christians the lower values of Chi Square may be attributed to the smaller number of respondents in the second, third and fourth categories, that is, religion important on family occasions, when help is needed, and never. In all, only 18 out of 70 Christian respondents opted for these three choices, and only two were in the family occasions category, that is, 3 per cent, whereas for Hindus the corresponding proportion opting for family occasions was 20 per cent.

However, in spite of poorer statistics for Christian respondents Chi Square for questions 3, 4 and 5 is sufficiently high to establish that the third hypothesis – that the importance attached to religion is inversely proportional to the degree of perceived conflict between science and religion – is valid for both Hindus and Christians considered as separate groups.

The influence of science

The purpose of the interviews was to clarify the questionnaire responses, to investigate further the manner in which science affects religious belief (both negatively and positively), and to explore in more detail the issues raised in the nineteenth and early twentieth centuries following the introduction of Western science into India. We shall not discuss any further the clarification of questionnaire responses; in this section we consider the influence of science on belief. In this and subsequent sections the religious community, and, in some cases, sub-community, will precede the respondent's code number as explained in Table 7.1. One hundred and fifty-five interviews were carried out by the investigator, 105 at the Indian Institute of Science, Bangalore, and the remainder at the other three centres.

A surprisingly large number of respondents were not only emphatic that science had changed their outlook, but could state almost the exact day on which they abandoned their religious beliefs:

> At University College, Tirupati, we learned the arguments for God's existence. According to A. J. Ayer they are linguistically invalid.... Until 1970 I believed religion – until a few months ago, in fact. (Hindu, 2-480)

> I abandoned religion one year ago after approximately one year of doubts. At Madras Christian College I was quite devout.... My loss of belief was partly due to reading Bertrand Russell, Ayn Rand and J. Krishnamurthi. Science gives definite conclusions, theology is based on imagination and guesswork. (Hindu, Sri Vaishnavite, 2-481)

> I stopped believing religion one year ago ... I realized that religion is based on blind facts. It arrests scientific progress and harms one's own personal development. (Hindu, 2-476)

> I lost my faith through studying science. During my first year at the Institute I went twice every week with my friends to the temple. I used to worry about the way some of them prayed only before exams – once a friend prayed in an autorickshaw for a train reservation.... Suddenly eight months ago all these things came together and I abandoned my religion. (Hindu, 2-471)

> I gave up religion when I was nineteen.... Astronomy completely destroys the old Hindu world-view. (Hindu, Saraswat, 2-477)

Two Muslims described their rejection of religion as follows:

> I began to doubt Islam...and I talked about my doubts to Hindu and Christian friends. My parents taught me that God predestines and punishes. There was no latitude for change due to the human will. My second phase of rejection came at high school where I discovered atheistic Marxism.... In Islam knowledge is built on assertions.... Some Muslims can argue that scientific evolution proceeds by the will of Allah, but basically Genesis is unquestioningly authoritative. (Muslim, Sunni, 2-459)

> I began to feel religious tension at University College, Trivandrum. Science played a part in the rejection of religion.... Many Muslims hold together the givenness of Islam and, say, Darwinism. They deny a contradiction, but I cannot do this and must reject religion. There is also an unresolved conflict associated with the Hindu view of karma in Islam – God and his will, but we can and must act.... The process of rejection was complete two years ago. (Muslim, Sunni, 2-451)

The second respondent summed up very clearly two problems frequently referred to by Muslims – Darwinism, and the conflict between human activity and the will of God.

Respondents who claimed that the study of science had increased their religious beliefs were enthusiastic but seldom as detailed about their reasons as those just quoted. The following statements are fairly typical:

> In some respects science has been the light of my life. (Hindu, Sri Vaishnavite, 2-449)
>
> Science has increased my beliefs and given me a sense of wonder. (Hindu, Iyer, 2-396)
>
> [My degree course] has given me a more clear understanding of both science and religion. (Hindu, Sourashtra, 3-1622)
>
> My religion is giving me more knowledge about science with the further studies because the Holy Book contains all sorts of science. (Muslim, 3-1624)

According to a Mirandian at Delhi University:

> I used to follow religion blindly. Science has given a better view of some things. Religion means a lot to me although I no longer hold all the religious opinions. I don't believe in the Hindu stories of creation and all that. (Hindu, Ramakrishna Mission, 1-4226)

Another Mirandian claimed that on the whole 'science is helping religious belief' (Christian, Methodist, 1-4225).

Asked what influence science had made on his understanding of the relationship between science and religion, a scientist from Kolkata at the IISc explained:

> Initially I thought that science was to do with material things and religion with spiritual things. Now I believe that the goal is the same or similar.... God is the goal towards which the whole of humanity is converging. (Hindu, 2-587)

All the respondents from which the previous quotations have been taken admitted to having been influenced in their beliefs either positively or negatively by science. But there were others who, although maintaining that science had not affected their views at all, conveyed a degree of uneasiness which suggested that they were beginning to feel a certain amount of inner tension. An IISc scientist from Kerala remarked: 'I do not probe the relation between religion and science. Religion does not conflict with science' (Hindu, Iyer, 2-462).

A biochemistry research scientist at the IISc became quite angry when the topic of evolution was mentioned: 'What evidence exists for it? Darwin may be completely wrong' (Hindu, Madhva brahmin, 2-475). To support this contention he then produced a paperback by the Union of Evangelical Christian Students of India which set out to prove the literal truth of Genesis![13] Another IISc scientist who first completed a questionnaire anonymously, and then a second one with his name on it, explained: 'How can I put into writing my doubts about the beliefs of my parents?' (Muslim, Shī'ite, 2-419). A science undergraduate at Kirioli Mal College, Delhi University, contended: 'It is claimed that the Americans have gone to the Moon, but according to my priest and our mythology they will find they are wrong' (Jain, 1-1022).

There was a tendency for some students to avoid a conflict between science and religion by limiting both the boundaries of science and the scope of religion. This was true of Christians who belonged to conservative Evangelical groups. Asked whether science had influenced his beliefs, one Evangelical Union member replied: 'Science has made no change to my beliefs. My faith is independent of asking questions' (Christian, Church of South India, 2-461). On the subject of evolution another member of the same group at the IISc claimed:

> Darwinism looks proven and Genesis is not scientifically proven, so you can avoid a conflict. It is a matter of faith. (Christian, Church of South India, 2-463)

Respondents who belonged to reformed Hindu groups such as the Arya Samaj, Brahmo Samaj and the Ramakrishna Mission claimed that the teachings of these organizations were more compatible with a scientific outlook than those of traditional Hindu belief. A research physicist at the IISc stated that he had become interested in the Ramakrishna Mission because it took seriously the issues posed for religion by science: 'It has its faults but it is progressive; it is not entirely rational but answers you more' (Hindu, Ramakrishna Mission, 2-403). A Kirioli Mal undergraduate explained:

> I like the Ramakrishna Mission because it teaches the universality of religions. Also their library contains science text books and books about science and religion. (Hindu, Ramakrishna Mission, 1-1089)

More important than their actual membership of reformed Hindu groups was the fact that many respondents had been influenced by literature produced by them. Many possessed copies of Vivekananda's and Ramakrishna's works and several drew attention to Vivekananda's claim that whichever parts of religion did not stand up to scientific investigation ought to be discarded. Generally speaking, the Reformers were felt to be dealing with issues still posed by science, and their books were readily available and inexpensive.

It will be clear from what has been said in this section that the religious beliefs of many respondents had been influenced as a result of exposure to science. In some cases scientific ideas were associated with changes resulting in the complete abandonment of religious belief. In others the study of science led to an enhanced respect for religion. For some exposure to science seems to have been associated with a shift of doctrinal emphasis or a desire to limit the scope of both religious and scientific enquiry in order to avoid conflict.

Issues of historical significance

Issues which were of particular importance in the ferment created by science in the nineteenth century included Darwinism, scientific–historical approaches to scripture, belief in progress, and new ways of regarding the physical universe and its extent in space and time. In this section the responses to the questionnaire and results of the interviews will be considered in the light of a few of these issues.

The two questions in part B of the questionnaire primarily concerned with issues which were important historically were as follows:

4. Which, if any, of the following *do not* agree with a religious outlook?
   (a) Biological evolution
   (b) Theories of universe's origin
   (c) Technological progress
   (d) The use of reason
   (e) The necessity for proof
   (f) Other.........
5. Which, if any, of the following *do not* agree with a scientific outlook?
   (a) Existence of the soul
   (b) Prayer
   (c) Life after death
   (d) Reincarnation
   (e) Miracles
   (f) Other.........

All respondents who were interviewed were requested to clarify their attitude to biological evolution and reincarnation, and where time permitted they were also asked to explain their views on the other items. As has been explained, reincarnation was not an important issue historically, but it was included here because preliminary discussions indicated that there was uncertainty about it.

The results of a Chi Square test for responses to question 4a requesting each respondent to specify whether or not there is conflict between biological evolution and religion for three degrees and one degree of freedom were obtained. For three degrees of freedom (df = 3) Chi Square is never less than 5.10, corresponding to a level of statistical significance of better than 0.2.

The majority of respondents did not seriously question the biological theory of evolution. Those who did were mainly Evangelical Christians, and a small number of Muslims, some orthodox Hindus, and members of certain sects. A member of the Adi Sanatan Deity Religion doing research in biochemistry completely denied the possibility of evolution: 'A human soul can only go into a human body. Man can never have evolved from an ape. Darwin is wrong' (Adi Sanatan Deity religion, 2–473).[14] It is interesting that the crux of this respondent's objection was not that religious revelation is threatened by evolution, but that Darwinism presupposes a common ancestry for humans and animals.

The following statements reflect a range of attitudes:

> You can't prove that God exists...but it's important because you feel God exists and you can't accept that you're just another animal. (Hindu, Brahmo Samaj, 1-5317)

> There is a prejudice about the common origin of humans and animals but this is really no problem for Hindus. No supernatural being is responsible for an act of creation. (Hindu, 2-417)

> Hindu creation mythology is in the Purāṇas. There is no idea of evolution in Hinduism. Krishna existed first, then humans and animals co-existed

together – this idea does not conflict with Darwinism. Through rebirth we are basically the same as animals, though the best animal. (Hindu, 2-415)

Darwinism does not agree [with religion]. It is not mentioned in scripture that we were apes and I don't believe we were. (Muslim, 2-442)

Almost all the other interviewees who felt that there was conflict over biological evolution expressed it in terms of differences between scriptural and scientific types of explanation. The following statements are characteristic of the manner in which many respondents reinterpreted scripture:

Darwinism is very similar to the ten incarnations of Vishnu. (Hindu, Iyengar, 2-401)

Darwinism agrees with the Rāmāyaṇa: at the time of the war between Rama and Rāvaṇa there were two species, the monkeys and the *rākṣasas*. (Hindu, 1-5345)

Brahmā created – Darwin's theory applies at a later stage. (Hindu, Neogi brahmin, 2-398)

According to science the earth evolved. According to religion God created Manu and his girl, and then there was a flood. (Hindu, 1-5272)

According to Hinduism after a period of havoc Manu collected all specimens and planted them. Darwinism does not conflict with this general idea. (Hindu, Ramakrishna Mission, 2-421)

According to the Granth, God closed his eyes, and when he opened them the five elements were there. From these beginnings life evolved according to Darwin. (Sikh, 2-437)

According to Genesis God created. According to biology life evolved via unicellular animals. Biology doesn't agree with the Old Testament. I don't like the Old Testament. (Christian, Mar Thoma, 1-2119)

Two interviewees at the IISc, a Hindu Iyer and a Madhva brahmin, suggested that karma and reincarnation were alternative and rival processes to biological evolution. Another IISc respondent observed that whereas the Hindu tradition presupposes a cyclical time sequence, Darwinism is linear in its concept of development.

From these statements it will be clear that discussions in the interviews raised a number of problems directly related to the biological issues which were important in the nineteenth century – in particular the idea of a common ancestry for animals and human beings, and attitudes to scripture and religious revelation in general.

The results of the statistical test for responses to question 4b were also obtained. For three degrees of freedom Chi Square is large for both the Delhi and Bangalore samples, and in each case the level of significance is better than 0.001. Here, as elsewhere, low values of Chi Square for Kottayam and Madurai groups may be attributed to a small number of responses in categories other than religion is important 'at all times'. Some of the more thoughtful observations about this issue were as follows:

> The human mind is still searching for the origin of the universe. All is *māyā*.... Relativity shows that an observed fact in science may not be a real one. (Hindu, 2-474)

> According to Hinduism Brahmā creates all things at once.... This may be correct; even science says that the planets came from a huge star – perhaps this can be interpreted as Brahmā. The meson is the elementary building block of science – perhaps Brahmā is a huge meson. Everything converges to Brahmā – the Oneness. (Hindu, Smārtha brahmin, 2-468)

> According to the theory of relativity time is relative and one second may be thousands of years. This type of interpretation can overcome conflicts with Hindu theories. (Hindu, 2-393)

> As far as the cooling of the sun is concerned Hindus believe that the universe existed from infinite time and had no origin. (Hindu, Ramakrishna Mission, 2-421)

> Cosmology may be cyclic unless Fred Hoyle is correct. Let science decide. (Hindu, Iyengar, 2-427)

With the exception of the last respondent, a Homi Bhabha research fellow, the tendency was to adduce cosmology and relativity to support religion rather than to 'let science decide'.

The following statements are illustrative of the variety of responses which arose in the interviews over reincarnation and life after death (question 5, c and d):

> I feel that people do not always get what they deserve – some good people have a hard time as though something done in a previous existence might be responsible. (Hindu, Nambūdri brahmin, 2-388)

> Science permits reincarnation.... Religion does not permit such artificial scientific methods [as birth control]. Controlling birth may mean controlling someone's reincarnation. (Hindu, Arya Samaj, 1-5347)

> I believe in rebirth. In the Gītā, Krishna says, 'In every age I come back'. (Hindu, 2-439)

> Science training has modified my beliefs and the idea of rebirth has been discarded first. (Hindu, Ramakrishna Mission, 2-408)

> Reincarnation is not possible; when you're dead, you're dead. (Hindu, 2-404)

None of the interviewees mentioned Radhakrishnan's attempts to adapt reincarnation. There were, however, several references to Vivekananda's teaching on the subject. One respondent, whose father seems to have been a full-time palmist and astrologer, claimed to have been particularly impressed by Vivekananda's works from an early age:

> At the age of fourteen I was challenged by an exposition of Vivekananda's teaching. Vivekananda has been very important to me ever since. [He says] 'The body will die, but I shall not die.' The idea of karma... in this passage... implies rebirth and the timing is related to the stars and planets. (Hindu, 2-414)

Several references were made to parapsychology during the interviews, and opinions were divided as to whether or not any scientific experiment could ever prove or disprove the theory of reincarnation. One respondent expressed interest in experiments in parapsychology, adding the comment: 'Personally I don't believe that one soul coupled with another body makes sense' (Hindu, Bhumiar brahmin, 2-425). This type of argument against belief in reincarnation is essentially the same as one of the major criticisms urged against rebirth by the Cārvākas.[15]

The choice of authorities quoted by respondents on the questionnaires and during the interviews is an important indicator as to the manner in which they themselves were responding to a process of secularization. Many Hindu respondents frequently quoted the opinions of the leaders of the nineteenth-century reform movements, particularly Vivekananda and Ramakrishna. The former was praised by one respondent for his ability to see the universe 'as a whole', and by another for his religious universalism: 'Vivekananda... is the ideal Hindu because he can live with and accept other religions' (Hindu, Smārtha brahmin, 2-390).

One research scientist at the IISc paid regular visits to the Swami Ramakrishna Ashram:

> I go twice a year to the Swami Ramakrishna Ashram to stay two weeks at a time.... Science has increased my interest in religion.... Religious and scientific approaches run parallel.... I have been influenced by Vivekananda. (Hindu, Iyer, 2-399)

Among more contemporary thinkers former president of India and Oxford philosopher S. Radhakrishnan was often mentioned, but very few respondents seemed to have read his books. Publications by the Bharatiya Vidya Bhavan were extremely popular, and articles about science and religion in *Bhavan's Journal* were frequently referred to during the interviews. Swami Ranganathananda's *Message of the Upanishads*, also published by the Bharatiya Vidya Bhavan, was in circulation among a section of students and lecturers at Delhi University.

Commentaries on the Gītā were very much in evidence at the IISc, but most of them were in Tamil or Malayalam. One respondent was particularly fond of Gandhi's *Discourses on the Gītā*, but none the less criticized the Gītā for inconsistencies:

> I read the Gītā every day – the Gītā is a dialogue between the *ātman* and the mind; it never occurred on any war field. There are conflicts within the Gītā;

[it contains the view that] there are certain days when you may die – this suggests an element of luck – your duty and how you have performed it are more significant. I believe that dying on an inauspicious day doesn't have bad effects. (Hindu, Iyengar, 2-401)

Few references were made to Dayanand Saraswati in spite of the fact that his books were as cheaply and readily available as those of Vivekananda. Muslims and Christians seldom mentioned authorities other than the Qur'ān and the Bible, though a few of the former had read Iqbal and Humayun Kabir.

Discussion of the results and conclusions of investigation

The results of this investigation were broadly in agreement with the findings of D. L. Jayasuriya and Surajit Sinha mentioned earlier.[16] The sample groups at the four major centres visited consisted of younger scientists than those interviewed by these investigators, but age was not found to be an important factor in relation to basic attitudes towards religion, and much of the interview data which has been quoted was obtained from older PhD research scientists and junior members of staff at the IISc.

The three initial hypotheses which were framed at the outset of this investigation were designed to clarify the type of relationship which might exist between science and religion, and the impact, if any, of science on religious belief. The third hypothesis, namely, that the degree of perceived conflict between science and religion has an inverse correlation with the importance attached to religion, was found to be valid. The first two needed modification, and it is suggested that from the point of view of further research they might be reformulated as follows:

Hypothesis (a) The study of science can either strengthen or reduce the religious beliefs of individual scientists, but the overall tendency is to diminish them.

Hypothesis (b) The greater the extent of scientific commitment on the part of scientists the greater the felt need on their part to relate their scientific and religious beliefs.

It was not found necessary to modify the hypotheses for different religious traditions, and the same methods were used consistently throughout the investigation. The initial assumption that the term religion would tend to possess a generally accepted meaning among groups of scientists living in similar environments seems to have been justified. Although the primary purpose of the investigation was to assess the extent to which scientists were able to relate science and religion and to validate the hypotheses, quite a lot of subsidiary information was gathered, and much of it would appear to indicate a very complex interrelation of social and religious factors.

Attention has been drawn to the abrupt manner in which some respondents abandoned their previously held beliefs. In some respects this was similar to the

128 Beliefs of Indian scientists

phenomenon of conversion as encountered in the Abrahamic traditions, though in a reverse direction. In many cases emotional commitment was preceded by a period of intense reading and followed by a sense of isolation. The following statement by a respondent at the Indian Institute of Technology in Delhi brings out the complexity of the changes brought about by a new scientific environment:

> At school we had religion as something to fall back on. It was a Catholic school, but we Hindus used to think of God as the Lord Śiva – that was the image we thought in terms of. During the vacations at home we celebrated the festivals regularly. Now I go home very seldom – we can stay on and work in the vacations here. I have come to think of religion as love, kindness or affection. The feeling I had of Śiva at school has gone.... God is *shakti*, which to me symbolizes energy and goodness. I don't think science has really played any part in this transformation. It is more the new environment and the influence of new forces – in part scientific perhaps. (Hindu, 1-5348)

In contrast to the nineteenth century when Western scientific ideas first entered India on a large scale some important Hindu concepts were no longer felt to be adequate. In particular some Hindu respondents had misgivings about belief in reincarnation and karma. In the nineteenth century neither of these doctrines was felt to be vulnerable to scientific advances, and it could be argued that the reason why Darwinism never posed a problem for educated Hindus was that through the doctrine of reincarnation the common origin of humans and animals was perfectly natural.

At the same time some of the scientific issues which troubled nineteenth-century educated Hindus continue to create problems for religious belief today. Many Christians and Muslims were unable to reconcile Darwinism with scripture, and some Hindus went to great lengths to demonstrate similarities between Brahmā's creation, Vishnu's incarnations and scientific theories of evolution. Few respondents seemed sufficiently well informed to argue about quantum theory or relativity, but there was some interest in cosmology among scientists at the IISc. The demand for rationalism and proof was frequently made, often without consideration for whether or not a type of proof used in a particular scientific discipline was legitimate when transferred to another. In some cases failure to fully understand the methods of science produced unnecessary conflict with religious beliefs – a situation paralleled in the nineteenth century.

The major findings of the questionnaire and interview investigation may be summarized as follows:

1 The popular idea that scientists keep their scientific and religious beliefs completely separate and never consciously try to relate them is not correct. Between 60 and 90 per cent of respondents at all four centres visited maintained that there is some relationship between the two, the nature of the relationship as indicated by the investigation being that the degree of conflict between science and religion in specific areas can be inversely correlated with the importance attached to religion (hypothesis 3).

2   While it may not generally be true that science is a cause of conflict between science and religion, it is often associated along with other factors with changes in belief. Quite often the change is one in which religion is given less importance and sometimes rejected completely. There is, however, a significant proportion of scientists whose beliefs have been strengthened by the study of science.
3   A superficial grasp of science can be the cause of more conflict between religion and science than a more mature understanding.
4   Several of the areas which played an important role historically continue to exercise scientists today. Darwinism and scientific–historical approaches to scripture still raise doubts, and reincarnation, which was not problematic historically, seems to be becoming increasingly so today. There was little awareness of the possibility that quantum theory and relativity might raise issues relating to religious belief, and nobody mentioned the Einstein–Tagore conversations.
5   In the same way that the influence of science cannot be entirely isolated from the consequences of other factors, so religious beliefs and social characteristics seem to be bound up in a complex manner. Thus, for example, the majority of those for whom religion was important on family occasions could not be described as predominantly rural, southern or anything else. While this investigation was not directly concerned with these matters, it did appear that the interactions between modern and traditional religious and social patterns are extremely complex.

8 How clear is reason's stream?

Tagore's beliefs will be considered in some detail together with Einstein's understanding of religion, the dialogue that took place between them on two occasions in 1930, and its ongoing implications from a variety of perspectives.

Writing in *The Guardian* newspaper in the autumn of 2005, the editor of a management journal recalls hearing village schoolchildren in Bangladesh reciting Tagore's poems from memory 'with obvious delight'. 'But', he adds,

> he was also a social development philosopher and social action researcher who undertook pioneering experiments to promote collective grassroots initiatives in rural West Bengal in the 1920s.... His emphasis on human creativity, self-reliance and social action offers the basis of a participatory development paradigm. Tagore in many ways foreshadowed Amartya Sen's articulation of 'development as freedom'.[1]

Amartya Sen's 'welfare economics' has been described elsewhere, and we shall not try to evaluate all of Tagore's many-faceted achievements here.[2] But we shall study aspects of his thought which may bear on his understanding of science and his reactions to Einstein's arguments.

Tagore's major themes

In the prologue to Krishna Dutta and Andrew Robinson's determinative *Selected Letters of Rabindranath Tagore*, Amartya Sen summarizes Tagore's characteristic themes as nationalism, traditionalism, educational commitment, freedom of the mind and interpretational epistemology.[3] We shall discuss these in relation to Tagore's attitudes to science and religion.

Tagore's nationalism was consistent with his Brahmo lineage. Like Ram Mohan Roy he took *advaita* Vedānta to be the 'characteristic world view' of India and the 'basis of Indian unity'.[4] Like Roy also, he had a tendency to express his views one way when speaking in English to Westerners (which may have been the case in his conversations with Einstein), and another when speaking or writing in Bengali. But whereas Roy's monotheism when writing in English was probably heightened under the influence of Islam and Christianity, Tagore's came from other directions (Bāul devotionalism, for example).

Tagore was greatly influenced by his father, the redoubtable Debendranath, who succeeded Roy as leader of the Brahmo Samaj. Debendranath Tagore refocussed the Samaj on the Upanishads and forged *advaita* Vedānta into an anti-Western weapon. In essence his response to the West was to say: 'You westerners may be technologically superior to us, but we are more spiritual, and our spirituality is based not on your philosophical and religious dualism which separates God, humanity and nature, but on the One underlying the many – that is *Brahman*.' Thus when Einstein insisted to Rabindranath on the continued existence of objects independently of their human subjects, the poet may have responded in terms echoing his father's monistic spirituality.

Debendranath was uncomfortable about his son's friendship with the Roman Catholic Brahmabandhab Upadhyay – and, possibly, to a lesser extent, with C. F. Andrews. The third leader of the Brahmo Samaj, Keshub Sen, had moved much too far in the direction of Christian theism for Debendranath's liking, and the last thing he would have wanted was for his son to follow suit. Rāmānuja's qualified theism would have been more acceptable, but not Christianity or Islam, even though Rabindranath was extremely popular among Bengali Muslims.[5]

According to David Kopf, Rabindranath's nationalism was tempered from the opening years of the twentieth century by a growing sense of universalism. But in 1919, following the Jallianwala Bagh massacre of a defenceless crowd by British troops, Tagore surrendered his knighthood.[6] Amartya Sen notes Tagore's differences with Gandhi over the use of the spinning wheel (*charka*) and other traditional practices, accusing him of 'exploiting [the] ... irrational forces of credulity in our people'.[7] He also strongly disagreed with Gandhi that the 1934 Bihar earthquake was God's punishment for the sinful practice of untouchability – which Tagore also abhorred.

Tagore's views about science were generally positive, as the following examples indicate:

> Science may truly be described as mysticism in the realm of material knowledge. It helps us to go beyond appearances and reach the inner reality of things in principles which are abstractions; it emancipates our mind from the thraldom of the senses to the freedom of reason.[8]

> Though science brings our thoughts to the utmost limit of mind's territory it cannot transcend its own creation made of a harmony of logical symbols. In it the chick has come out of its shell, but not out of the definition of its own chickenhood. But in India it has been said by the *yogi* that through an intensive process of concentration and quietude our consciousness *does* reach that infinity where knowledge ceases to be knowledge, subject and object become one – a state of existence that cannot be defined.[9]

The sense of both these statements is *advaitic*, with science pointing in the direction of the One underlying the many, but the highest knowledge consists of a spirituality that defies definition. That spirituality transcends distinctions between subject and object (which is crucial for scientific experimentation) and incorporates 'profound love' for a Being who is the 'goal that lies immensely beyond all that is comprised in the past and in the present'.[10]

The God of humanity

Tagore's notion of the supreme Being is intimately bound up with his understanding of relationship as foundational to reality:

> Reality is not based in the substance of things but in the [principle]...of relationship. Truth is the infinite pursued by metaphysics; fact is the infinite pursued by science, while reality is the definition of the infinite which relates truth to the person. Reality is human; it is what we are conscious of, by which we are affected, that which we express. When we are intensely aware of it, we are aware of ourselves and it gives us delight. We live in it, we always widen its limits. Our arts and literature represent this creative activity which is fundamental in man.[11]

Applying this insight to science, Tagore bridges the gap between atomic particles and God in an original way:

> It is not merely the number of protons and electrons which represents the truth of an element; it is the mystery of their relationship which cannot be analysed. We are made conscious of this truth of relationship immediately within us in our love, in our joy; and from this experience of ours we have the right to say that the Supreme One, who relates all things, comprehends the universe, is all love – the love that is the highest truth being the most perfect relationship.[12]

Thus the Supreme One is the ultimate guarantor of all relationships, and is love. But in so far that we humans are conscious of this truth of relationship, the 'human spirit' and the 'magic touch of personality' can move beyond the realm of science:

> The details of reality must be studied in their differences by science, but it can never know the character of the grand unity of relationship pervading it, which can only be realized immediately by the human spirit.[13]

> That which merely gives us information can be explained in terms of measurement, but that which gives us joy cannot be explained by the facts of a mere grouping of atoms and molecules. Somewhere in the arrangement of this world there seems to be a great concern about giving us delight, which shows that, in the universe, over and above the meaning of matter and forces, there is a message conveyed through the magic touch of personality.[14]

But it is as a collective entity that humanity achieves its ideal:

> The nebulous idea of the divine essence condensed in my consciousness into a human realization. It is definite and finite at the same time, the Eternal Person manifested in all persons. It may be one of the numerous manifestations of God, the one in which is comprehended Man and his Universe.... Whatever character our theology may ascribe to him, in reality he is the infinite ideal of Man towards whom men move in their collective growth, with whom they seek their union of love as individuals, in whom they find their ideal of father, friend and beloved.[15]

Although Tagore's consistent use of non-inclusive language seems outdated, it must be remembered that many of his finest poems are written from a feminine perspective (parts of *Gītāñjali*, for example).

The movement of 'men' in 'collective growth' must be achieved in humility, reverence and wisdom. In a frequently quoted passage the poet makes the following announcement:

> The God of humanity has arrived at the gates of the ruined temple of the tribe. Though he has not yet found his altar, I ask [those]... of simple faith, wherever they may be in the world, to bring their offering of sacrifice to him, and to believe that it is far better to be wise and worshipful than to be clever and supercilious.[16]

Tagore's overall thesis is compatible with Darwinian evolution, to which he refers approvingly.[17] In relation to more traditionalist theories such as reincarnation, his approval is somewhat *sotto voce*; he occasionally refers to God as Vishnu (e.g. in *Gītāñjali*), and reinterprets *māyā* and *dharma* imaginatively.[18]

We have taken most of our quotations from the same source, *The Religion of Man*, published in 1931 at about the same time that Tagore met Einstein. Tagore did modify some of his views during his long life, and further details of these changes can be seen in Dutta and Robinson's *Selected Letters of Rabindranath Tagore*. Tagore's educational commitment lies primarily in his activities at Shantiniketan. When Vishva-Bhārati University opened in 1921, Tagore dedicated it to 'a common pursuit of truth, [where we can] share together our common heritage, and realize that artists in all parts of the world have created forms of beauty, scientists have discovered secrets of the universe, philosophers the problems of existence, [and] saints made the truth of the spiritual organic in their own lives... for all [humanity]'.[19]

Shantiniketan celebrated variety. Satyajit Ray's 'inclusive vision' was forged there; Indira Gandhi was a student for several months. 'Freedom of mind' – Amartya Sen's fourth major theme which he attributes to Tagore – is expressed most decisively in education, though all of Tagore's areas of concern should be seen in 'the light of his strong attachment to the importance of living and reasoning in freedom.... It is in the defence of freedom and fearless reasoning that we can find the lasting voice of Rabindranath Tagore.'[20] Tagore's tribute to the ideal homeland in *Gītāñjali* is too well known for detailed quotation ('Where the mind is without fear... where the clear stream of reason has not lost its way.... Into that heaven of freedom, my Father, let my country awake').[21]

Amartya Sen characterizes the fifth of Tagore's major themes as 'interpretational epistemology'. We quote his comments in detail because they bear directly on his view of science and his conversations with Einstein:

> While Tagore was totally opposed not only to ignoring modern science in trying to understand physical phenomena, and particularly critical of giving ethical failures a role in explaining natural catastrophes, his views on epistemology were interestingly heterodox. The report of his conversation with Einstein... brings out how insistent Tagore was in interpreting truth through observations and reflective concepts. In this framework, assertions about truth in the absence of anyone to observe or perceive or conceptualize it appeared to Tagore to be deeply problematic. When Einstein asks, 'If there were no human beings any more, the Apollo Belvedere no longer would be beautiful?', Tagore asserts, 'No.' Going further – and into much more interesting territory – Einstein says, 'I agree with regard to this conception of beauty, but not with regard to truth.' Tagore's response is: 'Why not? Truth is realised through men?'
>
> This is, alas, not the occasion to discuss this engaging issue further. We could ask for clarification as to the sense in which 'realisation' is being used. Some would compare Tagore's position with certain recent philosophical works on the nature of reality, particularly Hilary Putnam's argument that 'truth depends on conceptual schemes and it is nonetheless "real truth"'.[22] Tagore's speculations on these issues, which were invariably interesting, were never systematically followed.[23]

We shall consider these issues again presently.

Tagore's interest in science can be traced to his early teens. He loved astronomy and, when in Britain many years later, visited the Greenwich observatory. He knew Jagadish Chandra Bose well, and was persuaded by Meghnad Saha to write a book on science in Bengali – he dedicated it to S. N. Bose. He collaborated with P. C. Mahalanobis, statistician and physicist, who became General Secretary of Vishva-Bhārati University in 1921. His encounters with European scientists and scientifically minded philosophers such as Bertrand Russell were numerous – he met Arnold Sommerfeld and Werner Heisenberg in India. *The Golden Book of Tagore*, published in 1931, was sponsored by a group of distinguished figures who included J. C. Bose, Gandhi and Einstein.

Bertrand Russell appreciated Tagore's poems, but wished he could have read them in Bengali. After meeting Russell in 1912 Tagore joked of him and his associates, 'Such nice men; and they would so much like to believe in God, but they want a receipt first!' He saw parallels between Russell's view that the essence of religion lies in the subordination of the finite part of our life to the infinite part, and the Upanishads, and wrote to him to say so. Russell apparently backtracked over what he had said in deference to Wittgenstein, but then, in 1967, at the age of ninety-five, decided to print Tagore's letter. Maybe Tagore had touched the Achilles heel of Russell's humanity: Russell once declared that he liked mathematics 'largely because it is *not* human... because, like Spinoza's God, it won't love us in return'.[24]

Einstein's religion

It is not always appreciated that some of Einstein's statements were intended to be humorous or, at least, deeply ironic. He corresponded with children; we have noted earlier an exchange between himself and a young girl who couldn't do her maths homework. Recently discovered documents show that Einstein composed equations for a card trick, and make light of his teaching duties.[25] There were many occasions on which Einstein displayed a wry sense of humour, and for this reason we need not take all his statements at their face value.

Albert Einstein was born on 14 March 1879 in Ulm, the first of two children. His father was an electrical engineer and his mother came from a business family; she was a talented piano player. Although both parents were Jewish, neither was particularly religious. In 1880 the family moved to Munich, where young Albert attended his first school, a Roman Catholic one where he was the only Jew in a large class. Anti-Semitism was common: on one occasion a teacher brought a long nail to a lesson and told the students that with such nails had the Jews crucified Jesus Christ. At the age of nine young Albert moved to high school at a Munich Gymnasium, where the subjects which he liked most, science and mathematics, were accorded the lowest status.

Einstein mentions two significant childhood 'scientific experiences' in his 'Autobiographical Notes'. The first was when his father showed him a compass. He notes that 'something deeply hidden had to be behind things'.[26] The second was to do with the 'lucidity and certainty' of geometrical proofs that he first encountered at the age of twelve. On reading more scientific books he came to doubt the truth of stories in the Bible. He records:

> Out yonder there was this huge world, which exists independently of us human beings and which stands before us like a great, eternal riddle.... The road to this paradise was not as comfortable and alluring as the road to the religious paradise; but it has proved itself as trustworthy, and I have never regretted having chosen it.[27]

In later years Einstein acknowledged the influence of both the Talmud and Christianity on his understanding of religion. He refused to become barmitzvahed – which is not a necessary condition for membership of the Jewish community, though most Jews – even liberal ones – observe it. He possessed a pair of phylacteries, which the Gestapo confiscated in 1933.

According to Max Jammer – probably the most authoritative author on Einstein's religious views – 'next to physics, the philosophy of religion and the quest for spiritual truth was perhaps the chief preoccupation of Albert Einstein.'[28] The following quote indicates the persistence of his religious beliefs:

> Behind all the discernible concatenations [of nature], there remains something subtle, intangible and inexplicable. Veneration for this force beyond anything that we can comprehend is my religion.[29]

This is not the atheism which Richard Dawkins attributes to Einstein. Once, when denying atheism, Einstein compared himself to a child in a huge library of books in many languages: 'The child dimly suspects a mysterious order in the arrangement of the books but doesn't know what it is. That, it seems to me, is the attitude of even the most intelligent human being toward God.'[30]

Of the New Testament, Einstein had this to say: 'No one can read the Gospels without feeling the actual presence of Jesus. His personality pulsates in every word.'[31] Some of Einstein's most familiar observations about religion may have been intended more laconically than those who quote them might wish. Thus his reaction to quantum indeterminacy, 'Does *der Alte* ['the Old One'] play dice?', can hardly have been more than a tongue-in-cheek response – he was far too capable a scientist to let belief in God intrude upon good science. Max Jammer comments upon the other well-known quote about science and religion being inseparable as follows:

> Although Einstein epitomized his philosophy of religion by stating that 'science without religion is lame; religion without science is blind', he never based his religion on logical inferences from his scientific work.[32]

What if Einstein's hearer had responded: 'then isn't it better to be lame than blind?' Might he not have roared with laughter at the joke?

Einstein began writing about science and religion from the 1920s onwards. He defined the terms as follows: science is 'systematical thinking directed toward finding regulative connections between our sensual experiences'. Religion is 'concerned with [human attitudes]... toward nature at large, with the establishing of ideals for the individual and communal life, and with mutual human relationship'.[33] Two statements which further clarify Einstein's views on science and religion are as follows:

> The most beautiful experience we can have is the mysterious. It is the fundamental emotion which stands at the cradle of true art and true science.[34]

> I do not believe in the God of theology who rewards good and punishes evil. My God created laws that take care of that.[35]

The first of these statements reminds us of Einstein's great love of art, and especially music. Once after listening to Yehudi Menuhin play with the Berlin

Philharmonic Orchestra, Einstein embraced the talented young violinist and declared: 'Now I know there is a God in heaven!' Einstein had no time for what is generally understood as mysticism. The second statement about a 'personal' God who rewards good and punishes evil is indicative of Einstein's deep sense of justice. He loathed hypocrisy, double standards, racism and all that the Nazis represented under Hitler.

Spinoza's God

'I love that man', Einstein once declared of Baruch Spinoza (1632–77), the Jewish philosopher who was excommunicated from his community for – among other things – denying the immortality of the soul. Spinoza's first publication expounded Cartesian philosophy, but ventured into new territory by identifying God with the fundamental scientific laws of nature – thus excluding 'miracles' in so far as these are understood in terms of divine intervention into the natural order.

Spinoza's *Ethics* was too controversial to be published in his lifetime. Einstein may not have subscribed to all of Spinoza's views, but his understanding of the infinite attributes of God as thought and extension according to which a mode of one was the same as of the other would have appealed to him; it also has resonances with Hindu philosophy. Since the human mind and the human body are modes of thought and extension respectively, this implies that they are the same thing, considered under the attribute, first, of thought and, second, of extension. It also means that any physical entity, as a mode of extension, must be the same as some mode of thought – must 'have a mind'.

Does Spinoza's argument mean that plants can feel pain as Jagadish Chandra Bose maintained? Possibly not, but it does imply that they have a quality of life not entirely dissimilar to ours, and that our minds die with our bodies – which is what upset the orthodox of Spinoza's day. In the last section of Spinoza's *Ethics* he claims that there is a part of the mind which survives bodily death, allowing a small space for human effort within what is otherwise strictly determined. Stuart Hampshire maintains that in this respect a significant strength of Spinoza's philosophy is that it is compatible with, and almost anticipates, current theories of evolution and genetics.[36]

Einstein also had much in common with Schleiermacher, who maintained that religion is not so much based on revelation or dogma, but on a 'feeling for the unity underlying all finite and temporal things'.[37] It is this feeling, Schleiermacher declared, that generates the idea of God:

> Our pious self-consciousness (*frommes Selbstbewusstsein*), by which we relate everything that stimulates or affects us to God, is identical with the recognition (*Einsicht*) that all this depends on, and is determined by, the unity of nature.[38]

Such a view is compatible with Einstein's cosmic religion; Albert Schweitzer held similar opinions. They are not too distant from the Vedānta of Shankara or Rāmānuja.

Meeting of minds

Albert Einstein and Rabindranath Tagore had their first recorded conversation in July 1930 at the former's home on a hilltop near Potsdam, not far from Berlin. Einstein, aged forty-two, came down the road to meet his gaunt, bearded guest, who later recalled about his host, 'His shock of white hair, his burning eyes, his warm manner impressed me with the human character of this man who dealt so abstractly with the laws of geometry and mathematics.' Commenting on Einstein's great simplicity, he added, 'There was nothing stiff about him – no intellectual aloofness. He seemed to me a man who valued human relationship and he showed toward me a real interest and understanding.'[39]

Einstein's opening question to Tagore, 'Do you believe in the divine as isolated from the world?', suggests that he had some familiarity with Hindu thought. Tagore's reply is central to his argument. Humanity is characterized by personhood which possesses infinite possibilities including our unique capacity to comprehend the universe. Nothing, however vast or minute, 'cannot be subsumed by the human personality'. Thus, Tagore argues, the truth of the universe is human truth.

Tagore proceeds to illustrate this assertion with a scientific example. There are gaps between the protons and electrons which constitute matter, but from our point of view it seems to be solid. (Tagore made this statement well before the complex interrelationships underlying fundamental particles were known.) Similarly, although humanity is made up of individuals, it is the interconnections of human relationships that give solidarity to our world. The entire universe, on account of its linkages with us, is a human universe. This argument has resonances with versions of the Anthropic Principle, though Tagore justifies it in terms of art, literature, 'and the religious consciousness of...[humanity]'.

Einstein sets out two conceptions of the nature of the universe: (1) the world is a unity dependent on humanity, and (2) the world is a reality independent of humanity. Before he can enlarge on the relationship between these two views, Tagore expresses his preference for the first, which he considers to lie in the realms of truth and beauty. Einstein requests clarification, which Tagore gives along the lines indicated in the first part of this chapter.

The sticking point in the conversation comes when Einstein asks what would happen if human life were to disappear. Would beauty still be beauty and truth still be truth? Tagore responds negatively to both. Einstein agrees with regard to beauty, but not with regard to truth – which he admits he cannot prove (because such a situation can never be susceptible to proof), but firmly believes.

Tagore elaborates his understanding of truth. It is 'the perfect comprehension of the Universal Mind'. Thus we move towards the truth via our experiences and 'illumined consciousness'. Einstein argues that, however arrived at, truth must be valid independently of humanity, citing the theorem of Pythagoras as an example. If there are realities independent of humanity, then there must be corresponding truths. If, as you say – he responds to Tagore – there is no reality independent of humanity, then how can there be any corresponding truth? Tagore sets out the Hindu view that absolute truth is *Brahman*, 'which cannot be conceived by the isolation of the individual mind or described by words, but can only be realized by completely merging the individual in its infinity'. This truth, Tagore explains, cannot belong to science, which can only ever explore the appearance of *Brahman*, that is, what appears to be true to the human mind.

At this point Tagore introduces the notion of *māyā*, which he translates as 'illusion'. We have already noted that this is an unhelpful translation, in that it gives the impression that a thing is not real, or is not really there, which is not what either Shankara or Rāmānuja meant by it. Furthermore, the notion of merging the individual into *Brahman*'s infinitude is alien to scientists for whom subject–object distinctions are integral to the scientific method.

More generally, Tagore's explication of Hindu philosophy in terms of *Brahman*, which cannot be described in words (i.e. objectively), but only through the union of the self with ultimate reality (i.e. subjectively), would be acceptable to both Shankara and Rāmānuja. But they would have disputed the nature of this reality. Rāmānuja levelled a series of arguments against Shankara's view of the world as appearance – he believed that his objective arguments were necessary in order to clear the way to ultimate liberation through subjective communion with Vishnu Nārāyana. For Rāmānuja there is always some differentiation between our individuality and God's – we are the body of *Brahman*, giving both us and the material world more value than that accorded by Shankara. Tagore probably felt that this would be unfamiliar territory for Einstein.

Einstein responds by fastening on Tagore's emphasis on collective humanity. Without denying it, he draws attention to everyday situations in which, for example, a table continues to exist in a house when there is nobody there to perceive it. Tagore agrees: the table continues to exist in isolation from individual consciousness, but if the universal mind were to disappear, then so would the table. Einstein insists that truth possesses a super-human objectivity: 'It is indispensable for us, the reality which is independent of our existence and our experience and our mind.' In extending the boundaries of truth beyond human collectivity, Einstein reflects his personal belief in cosmic religion. Which may be why, after Tagore's final attempt at clarification, Einstein retorts, 'Then I am more religious than you are!'

Tagore's closing statements do not carry his earlier arguments forward. He considers Einstein's table both as a macroscopic and as a microscopic object, conceding that we need to reconcile 'an eternal conflict between the universal human mind and the same mind confined in the individual. The perpetual process of reconciliation is being carried on in our science and philosophy, and in our ethics.' He concludes: 'If there be some truth which has no sensuous or rational relation to the human mind it will ever remain as nothing so long as we remain human beings.... My religion is the reconciliation of the Super-personal Man, the Universal human spirit, in my own individual being.'

The second conversation between Einstein and Tagore, which occurred on 19 August 1930, also at the former's house, was only partially recorded. The extract in Appendix B is too brief for any significant conclusions to be drawn. If the first discussion consists mainly of Einstein listening politely while Tagore sets out his Hindu philosophical view, the second extract has Einstein explaining his commitment to causality to an attentive Tagore. Tagore feels inspired by 'the constant harmony of chance and determination' which makes our otherwise desultory existence 'eternally new and living'.

Dialogue in context

At the time of the first recorded conversation with Einstein, Tagore had just concluded the Hibbert lectures in Oxford, published as *The Religion of Man*.[40] We have quoted from this book earlier in this chapter because it has more to say about science than Tagore's other publications, and because it gives a good idea about where Tagore was in his thinking when he met Einstein. He modified some of his views later in life.

It was noted in an earlier chapter that towards the end of *The Religion of Man*, which contains the first recorded conversation with Einstein as an appendix, Tagore 'distanced' himself from *advaita* Vedānta. He had been lecturing to an Oxford audience who would have warmed to the notion of those who 'have felt a profound love...for God, who is...the goal that lies immensely beyond all that is comprised in the past and the present'.[41] But with Einstein he expressed no such sentiments about God, even though Einstein opens the conversation with a reference to 'Divinity'.

At this point there is a problem, because the version of the discussion contained in the appendix to *The Religion of Man* differs from the version published in the *New York Times* (which Einstein approved). Here Tagore's response to Einstein's two views about the nature of the universe is as follows:

> *Tagore:* This world is a human world – the scientific view of it is also that of the scientific man. Therefore, the world apart from us does not exist; it is a relative world, depending for its reality upon our consciousness.

This is a more clear-cut statement of Tagore's position than the earlier one, and we prefer it rather than the more confused version in Appendix A.

If Tagore was coming to this conversation fresh from Oxford, what were the events and ideas which had been shaping Einstein's thinking prior to the meeting? Einstein began his scientific career as a positivist, that is to say he believed that theories are not only justified on the basis of facts from observations, but have meaning only in so far as they can be so derived. During the 1920s he embraced realism, which, in relation to science, is the view that most of the entities which are postulated in a scientific theory to explain what happens in a laboratory are real, independently existing things. Epistemological realism (i.e. realism that inquires into the possibility and nature of knowledge) is the view that a mind-independent world exists and that in perception we mentally grasp qualities and objects that are part of that world. Einstein appears to have moved towards this position following the success of his theory of General Relativity.

For Einstein the realist, the table in the house had to continue to exist when nobody was there. Tagore did not deny the continued existence of the table,

How clear is reason's stream? 141

but maintained that its existence must be perceived by a conscious human mind if it is to mean anything. He also claimed that although we conceive of our consciousness as individual, it has a universal character such that we can speak of the 'universal mind'.

What did Tagore mean by the 'universal mind'? He once compared the universe to a cobweb in which human minds are like spiders – 'for the mind is one as well as many'.[42] Dipankar Home and Andrew Robinson quote the physicist Max Born's observation that 'All religions, philosophies, and sciences have been evolved for the purposes of expanding the ego to the wider community that "we" represent.' They continue:

> If mind/consciousness, the first-person perspective, is somehow to be incorporated into physics, as certain physicists believe it should be, this would entail consequences as dramatic as those involved in the introduction of relativity by Einstein, for it would mean an acceptance that 'the lawfulness of events, such as unveils itself more or less clearly in inorganic nature' may, at least in principle, 'cease to function in front of the activities in our brain' – to answer Einstein's sceptical question addressed to Tagore in the affirmative. But Einstein could never accept this.[43]

The quote cited here is from *The Golden Book of Tagore*, and is not part of the Einstein–Tagore conversation. Had it been, then the ensuing discussion might have been more fruitful.

In our summary of Tagore's major themes based on Amartya Sen's analysis, it was noted that Tagore's philosophical position has similarities with that of Hilary Putnam, and, we might add, Thomas Nagel. In *The Many Faces of Realism* Putnam rejects 'local realism' (i.e. the notion that an atomic particle can be described in classical terms) which does not take into account the role of our minds.[44] This calls in question the distinction usually made between objective and subjective views of truth by inserting mind into reality. Thomas Nagel associates both the self and our objective viewpoint with human consciousness.[45] Roger Penrose has considered whether or not consciousness is a quantum phenomenon.[46]

Another view of science

We have said enough to indicate that the conversations between Einstein and Tagore cannot be characterized merely as an encounter between objective Western science and subjective eastern religion. In this section we shall consider another view of science, namely, that of Niels Bohr and the Copenhagen School, and how this might relate to Tagore.

Einstein's favourite example of philosophical realism was to invite his hearers to consider whether or not the moon continues to exist if we don't look at it. The physical world is objectively real, and this must include atomic and molecular particles because we can construct experiments that prove their existence.

Furthermore, as Einstein pointed out to Tagore, propositions such as the Theorem of Pythagoras continue to be true irrespective of human perceptions of them.

But according to Werner Heisenberg, whose Uncertainty Principle shook the scientific establishment in 1927, 'The laws of nature which we formulate mathematically in quantum theory deal no longer with the elementary particles themselves but with our knowledge of the particles.'[47] This view has something in common with positivism, and has been described as quasi-positivism. It is also subjectivist in that it remains agnostic about the behaviour and properties of the physical world until they have been measured.

There are several interpretations of quantum theory. The dominant one is that quantum probabilities become determinate on measurement – the wave function collapses when classical measuring apparatus is brought to bear. There are problems with this school of thought: it presupposes an indeterminate quantum world and a determinate classical world; it also raises questions about what kind of measuring instruments collapsed wave functions before humans evolved on this planet. Some physicists have identified classical measuring apparatus with consciousness. In which case we need to ask what kind of consciousness do animals possess? Is Schrödinger's poor cat conscious enough to determine the outcome of the experiment to which it was subjected?

Lawrence Osborn postulates a transcendent world observer, 'a divine mind whose observations collapse the wave functions on our behalf. In effect this would be the quantum-mechanical version of Bishop Berkeley's idealism.'[48] But then why should anything be left indeterminate by this observer for us to determine by our measurements? There is also a neo-realist or hidden-variable view of quantum mechanics which maintains that the statistical aspect of quantum mechanics only applies to groups of particles. Thus David Bohm distinguishes between the quantum particle and a guiding wave that controls what it does. There is even a 'many worlds' class of interpretations according to which Schrödinger's cat dies in one world and lives in another!

None of these interpretations of quantum theory would have appealed to Tagore, though their general features represent an alternative approach to Einstein's position. We have already noted the views of Putnam, Nagel and Penrose, of which Penrose's would probably have been the most acceptable to Tagore. Penrose suggests that our brains use quantum collapse to solve problems non-algorithmically (i.e. not along the lines of a computer, because computers are incapable of intuition). He associates this process with what he calls microtubules that occur within cells. These enable quantum effects within them to coordinate across groups of neurons to facilitate intuition.[49] This brings us back to the Hindu view of intuitive knowledge and may turn out to be a fruitful way forward for discussions between scientists and Hindu philosophers.

From the point of view of the conversations between Einstein and Tagore, it will be clear that Einstein represented one of two major streams of scientific thought – still valid; the other, represented by Niels Bohr and his Copenhagen associates, is arguably closer to Tagore's beliefs than to those of Einstein. But the aspects of Tagore's thought relating to universal humanity are beyond the scope of both scientific viewpoints at that time.

Contemporary perspectives

To what extent can contemporary Hindu, Muslim and other scientists identify with the above arguments and their proponents? A few cases will be considered in this section, beginning with a Muslim.

Dr Usama Hasan is senior lecturer in artificial intelligence at Middlesex University, and *Imām* of Masjid Tawhid in Leyton. He has studied theoretical physics at Cambridge University, where he gave a lecture on 'God and the New Physics' early in 2006. He commented on religion and science from a Muslim perspective as follows:

> As a student in Cambridge I found a tension between what is taught [about science] in the university and in the mosques. A Muslim friend in south Asia was so concerned about this that he vowed he would never let his children study physics or mathematics. But I am familiar with several contemporary writers on science and religion: Paul Davies, Stephen Hawking, John Polkinghorne and Richard Dawkins. Dawkins is right to attack fundamentalist Christians and Muslims; this is a reaction against people who misuse religion and is legitimate criticism. In response we must come to a deeper and more authentic religion.

In relation to the Einstein/Tagore discussion, Usama Hasan's sympathies lay more with Tagore:

> Einstein seems not to grasp Tagore's understanding about the mutual relationship between man as human being and Universal Man as the world. Hence he posits a disconnection between 'man' and 'everything else'. He mentions Pythagoras's theorem, which is indeed a mathematical truth of great beauty that exists independent of man.... However, that same principle of 'beautiful truth' that reverberates throughout the universe is also found within man, in both his physical and metaphysical constitution.

In relation to Tagore's interpretation of *Brahman*, which, as we have noted, would be acceptable to both Shankara and Rāmānuja, Hasan had this to say:

> *Brahman* here would seem to be very similar to the Qur'ānic Divine Name *al-Haqq* (truth or reality). There is a process involved in 'completely merging the individual in its infinity', and that is to follow the path of the exoteric Sacred Law with the esoteric Spirit. In a sense, this is the entire purpose of human life: to follow the Qur'ānic 'Straight Path'.... Note that *Brahman* is also known as, or related to, 'Brahma' which is very close to 'Abraham', and in fact seems to have the same four major consonants in the same order (BRHM).

> This is a striking feature which suggests a major link between the three Abrahamic religions and the Indian tradition.

Towards the end of the discussion, where Einstein refers to the problem of whether or not truth is independent of our consciousness, Hasan is in agreement with Tagore:

> I entirely agree with Tagore. 'Super-personal man' would seem to be the same as 'Universal Man'.... The all-pervading consciousness is related to the supreme Spirit, the Qur'ānic *Ruh* which is one of the greatest created entities. Just as an individual human being has a body and soul, Universal Man is the physical world with its supreme Spirit. This is how 'everything in the heavens and the earth sings the praises of God' – a truth found in the Bible as well as in the Qur'ān.

Hasan ends his assessment of the discussion by observing that 'the concluding statement of each figure says it all. Einstein is an exoterist who emphasizes transcendence; Tagore is an esoterist who emphasizes immanence.'

More generally, most Muslim writers in academic publications adopt positive attitudes to science. *The Cambridge History of Islam*, Volume 2, contains a substantial section about the history of science in Islamic societies, and concludes: 'Islam, of itself, did not offer any kind of opposition to scientific research, in fact quite the contrary.'[50]

Articles appear periodically about science and Islam by scholars such as Muzaffar Iqbal (originally from Lahore) and Syed Hossain Nasr. Writing in the *New Scientist*, Ziauddin Sardar distinguishes between a 'toxic combination of religious literature and "science"' which he compares to Creationism and Intelligent Design in the West, and a more mature and reflective strand which he calls mystical fundamentalism: 'Islamic science becoming the study of the "essence" of things'. He continues:

> The material universe is investigated as an integral and subordinate part of higher levels of existence, consciousness and modes of knowing. So science becomes not a problem-solving enterprise or objective enquiry but a mystical quest to understand the Absolute.[51]

Much the same could be said in relation to the other major Abrahamic traditions, Judaism and Christianity.

Dr Jacob Cherian, vice-principal of St Stephen's College, Delhi, a physicist and a Christian (Church of North India), sympathizes more with Einstein than with Tagore: 'The human concept of the universe is so different from place to place that no one concept can be called human. Some of these concepts are mutually exclusive.' Citing St Paul's letter to the Romans (chapter 2, verses 18–32), he maintains that people reject the notion of God in order to avoid accountability for their actions.

We consider four observations about the Einstein/Tagore discussion by Hindu science graduates from Delhi University (St Stephen's College), whom I taught in the mid-1990s.

Dr Supriya Goyal was born in Meerut in Uttar Pradesh. Her father was an officer in the Indian Army, her mother a professor of zoology. Of her religious beliefs she says: 'I was born and raised a Hindu; over the years I have remained a practising Hindu, but I consider myself to be more spiritual than religious.'

After graduating from St Stephen's College in chemistry, Supriya did her master's degree in chemistry at the Indian Institute of Technology, Kanpur. She has just completed a doctorate in materials science at the Berkeley campus of the University of California working on far infrared detectors for applications in space astronomy. Was her choice of research influenced by her beliefs as a Hindu?

> Not directly. But since my religious beliefs have played a role in shaping me as a person, I believe that indirectly they affect everything that concerns me. I believe in the Hindu doctrine that living beings have to live many lives and undergo many experiences before becoming one with the Divine.

Of the Einstein/Tagore discussion, she had this to say:

> I tend to agree with Tagore that even if absolute truth existed that was unrelated to humanity, it would be inaccessible to the human mind. Being both a scientist and a deeply spiritual person, I can understand both positions, but I identify more with Tagore. The relationship between science and religious philosophy is aptly summed up by Tagore when he says, 'In the apprehension of truth there is an eternal conflict between the universal human mind and the same mind confined in the individual. The perpetual process of reconciliation is being carried on in our science and philosophy, and in our ethics.'

Dr Bhismadev Chakrabarti is the son of an engineer working for Tata Steel; the family are brahmins, originally from West Bengal. Bhismadev graduated in chemistry from St Stephen's College, scored a first at Trinity College, Cambridge, and has recently completed his doctorate in psychiatry and experimental psychology. He is a brilliant Indian classical musician.

Bhismadev's religious beliefs do not influence his scientific work:

> My religious beliefs derive primarily from the *advaita* tradition in Hindu philosophy, which does not pose any contradiction to the scientific method. According to my beliefs there is one eternal life that we are all part of. Hence questions of reincarnation and resurrection become very proximate, and therefore irrelevant. To ask about these things is like asking a cell in one's nail, when it is not yet 'dead', if it believes in reincarnation – when all of creation represents a living whole.

In relation to the Einstein/Tagore discussion, Bhismadev, like Supriya, has more sympathy for Tagore's position:

> The conversation is delightfully logical, especially where Tagore highlights the relativism inherent in the 'scientific' method. I find Tagore's belief in God – if that is the right word – more healthily sceptical than Einstein's belief in absolute reality. The discussion as a whole offers a good basis for a relationship between science and religious philosophy because it exposes the fundamental divide between people who believe in an absolute reality and those who do not. Which of these schools one belongs to will determine the nature of one's identification with one's science and/or religion.

Dr Ajatshatru Mehta is a Punjabi, one of whose parents is Sikh, the other Hindu. He attended a Sikh school, went to St Stephen's College to read chemistry, and from there via Cambridge to Ohio State University where he has just completed a doctorate on powder X-ray diffraction. He describes himself as spiritual, but 'I don't think I am very religious':

> I believe in including good things from every religion in my life. My spirituality influences my work indirectly. It helps me to stay focussed and positive whilst I'm working. I strongly believe in reincarnation; the belief has been strongly etched in my mind since childhood.

Of the Einstein/Tagore discussion Ajat had this to say:

> It's very hard to say whom I agree with more, Tagore or Einstein. I am a scientist and hence like more structured views and may incline more towards Einstein. Tagore does touch on it in the conversation when he talks about the impersonal nature of science and the more harmonious view of religion. But I guess science is universal, religion is not. Science tends to unite people and, even though religion is supposed to do that, in reality it doesn't. People today are divided by human walls of religious discrimination. In India the clash between Hindus and Muslims, the deaths of Christian missionaries; in the USA the whole evolution debate tends to show that religious beliefs can cause social unrest. The views of Tagore point towards a utopian belief. In today's modern world very few people practise such religious ideas.
>
> The discussion is quite intellectually stimulating. I find it a bit perplexing that in the Western world if I utter the words 'Oh God' I am confronted with questions like: Being a scientist how can I believe in God? What god do I believe in? My answer as always is that science has not proved or disproved the existence of God. People are eager to believe that there is another planet like earth in a far-off galaxy but have a hard time accepting a God-fearing scientist. The whole evolution debate in the USA puts the topic of the relationship between science and religion into a new perspective. Both sides present quite convincing arguments, but trying to decide which is right is not very easy.

Lately I have started believing that science and religious philosophy are related. I have started to look at scientific developments in ancient civilizations. Ancient Indian civilization had holy texts about natural medications.... I guess the whole diversification of science and religion is quite recent.

Dr Ankur Barua was born in Assam. He describes his family as Vaishnavite, and his parents are both college lecturers. He graduated in physics from St Stephen's College with the highest first class degree of his year in Delhi University, read physics at Cambridge University, and has just completed a doctorate in comparative religion (mid-2006). He states that his religious views have no influence on his understanding of science, and he is agnostic about the possibility of post-mortem existence.

Ankur's comments about the Einstein/Tagore discussion are as follows:

The discussion (in 1930) took place during a 'happening' stage in the history of modern science when classical mechanics was gradually being revised by figures such as Heisenberg and Schrödinger, and by Einstein himself, though along independent lines. In retrospective terms, this interchange between a poet and a physicist could be read as a sign of things to come. In later decades, with the gradual waning of Anglo-American positivism, physics, literature, poetry, religion, metaphysics and philosophy would be drawn out of their respective spheres of influence (and their mutual indifference) into various forms of mutual engagement. At the heart of these inter-disciplinary confluences there stand out certain questions such as the nature of reality, the concept of truth, the deep structure of the universe and the possibility of value in a cosmos (allegedly) indifferent to human flourishing. These are also issues that Tagore and Einstein either directly address, or make passing reference to, during their conversation. Finally, the two figures can arguably be taken as fairly representative of two broad intellectual/religious streams: that of the classical science of post-Enlightenment Europe and that of the reformed Hinduism that emerged from the crucible of nineteenth-century Anglicized Bengal. However, this must not be viewed as an oppositional encounter between a scientist physicist and a religious poet: not only did Einstein frequently deal in his writings with issues broadly understood as religious, but Tagore belonged to a generation of Hindu figures who, in their highly creative and distinctive ways, had appropriated elements of European science into their thought. Consequently, one might expect that this dialogue will ultimately throw up, even if only implicitly, various meta-questions about the very nature of science and religion and their possible interconnections.

Asked to what extent he could identify with Einstein or Tagore, he responded:

As a student of physics who has now been involved in the area of religious studies for a significant period of time, I tend to view this conversation as, so to speak, a *non*-conversation. That is, Einstein and Tagore seem to be talking almost at cross-purposes: both latching onto two *distinct* concepts that are usually associated with truth. Einstein proposes a robustly realist conception of truth in terms of metaphysical realism, while Tagore tends to view truth more in terms of an individual's personal appropriation of reality. That is, for Einstein certain truths (for example, the square of the hypotenuse of a right-angled triangle is the sum of the squares of its other two sides) are such whether or not there are human observers, whereas Tagore seems to argue that these truths *become* true for individual centres of creativity who real-*ize* them in their personal being. Tagore distinguishes between two types (or even levels) of truth: the impersonal, mathematical, analytic, atomic, abstract truth of *science*, and the personal, intuitive, organic, holistic, integral truth of *religion*. He argues that different human beings 'cut up' reality in their distinctive ways, so that the scientific perspective is *one*, out of many possible, that is (circularly) adopted by the scientific man. In contrast to the empirical truths that are available at the scientific level stands the truth of ultimate reality which can be grasped not through a dispassionate stance (which allegedly is the methodological orientation of the scientist) but only through a personal *in*-volvement with *That* which is sought to be known.

Commenting on the dichotomies that Tagore proposes between science and religion, he made two observations, as follows:

First, a number of Wittgensteinian philosophers of religion have put forward a somewhat similar distinction between the subjective truths involved with religious perceptions and the objective truths from a scientific provenance. Their critics, even while admitting that they are correct in pointing out that religious experiences cannot be adjudicated on the basis of external norms derived from the realm of scientific enquiry, have pressed the point that members of various religious traditions (especially Christianity and classical Buddhism, Hinduism and Jainism) have *themselves* attempted to establish the rationality and the objectivity of their truth-claims. We can make this point in the case of Hindu metaphysics by taking up Tagore's own example of the absolute truth of *Brahman* which cannot be objectively described in human words but known only through the subjective union of the self with reality.... The Christian traditions, for their part, are characterized by a long history of intertwinings between two types or levels, even when their mutual distinctiveness has often been asserted and sought to be maintained. These intercrossings can be summarized under the motto *credo ut intelligum*: I believe subjectively, in revealed truths, in order that I may arrive, even if eschatologically, at an objective understanding of them. In short, then, the alleged clear bifurcation between objective truth in science and subjective truth in religion is not supported by the data of the history of religions.

Second, as philosophers of science have pointed out in recent times, there is a crucial component of subjectivity at the very heart of the scientific enterprise. In the words of the well-known slogan, 'all data is theory-laden', which implies that more than one theory can explain the *same* set of data.

Consequently, the choice *between* these theories is not, strictly speaking, a scientific one and is, in fact, usually justified on the basis of certain subjective aesthetic factors such as symmetry, beauty, internal consistency and comprehensiveness.

Thus, although these two conceptions of truth, the subjective and the objective, far from being opposed to each other, are in fact integral to and mutually intertwined in the scientific enterprise; and although the notions of subjectivity and objectivity that are characteristic of scientific investigation cannot be directly translated without remainder into the religious domain, both nevertheless have resonances in a number of Abrahamic and Hindu religious traditions which revolve around the worship of a transcendent Creator.

The argument presented here is not that scientists and religious believers employ the terms 'subjective truth' and 'objective truth' in *precisely* the same fashion, but that both the scientific and the religious spheres are marked by complex sets of *internal* relationships between these two aspects of the enquiry into reality. And some of these relationships *may* reveal, under close and careful investigation, significant parallels to encourage and inspire scientists and believers to enter into a mutual (re)examination of the metaphysical presuppositions that underpin one another's conceptual systems...

On the basis of Tagore's presentation of the difference between religious truth and scientific truth, one could argue that religion provides an encompassing meta-empirical framework within which various levels and dimensions of empirical scientific results can be accommodated. This two-fold level of truths would *somewhat* parallel the Thomist distinction between salvifically higher revealed knowledge and salvifically lower natural knowledge: the latter, pertaining to the mundane world, can be discovered without the apparent help of revelation, while the former, concerning matters of redemption, is made available to human beings only, and specifically, through divine initiative.

We have quoted this Hindu scholar at length because his comments offer a profitable basis for a better understanding of the relationships between and within science and religion in an inter-faith context.

Conclusion

Having introduced Tagore in an earlier chapter, we followed Amartya Sen's analysis by summarizing his major themes in terms of nationalism, traditionalism, educational commitment, freedom of the mind and interpretational epistemology. By considering Tagore under these headings we were able to build up a picture of his understanding of science and religion – which was not always consistent because, like earlier members of the Brahmo Samaj, he tended to express his views one way when communicating in English to Westerners (including Einstein) and somewhat differently when speaking or writing in Bengali.

We considered Einstein's religious beliefs in terms of his unorthodox Jewish background, his admiration for Spinoza and his sense of irony – the implication being that not all his better-known 'one-off' statements need be taken at face value! We discussed the 1930 conversations between Einstein and Tagore, commenting on them in the light of our prior understanding of their backgrounds. We noted differences between the published versions of the two recorded conversations.

In a later section we considered how the Einstein/Tagore conversations might have proceeded had the scientific point of view been represented by Niels Bohr or one of his Copenhagen associates, who held different views of particle physics, instead of by Einstein. Finally, we invited representatives of a wider group of contemporary religious and philosophical perspectives to reflect on the first of the Einstein/Tagore discussions. We concluded that there are sound reasons for situating what has hitherto been a fairly narrow discussion of the relationship between science and religion in a much broader inter-faith context which takes into account the internal debates within each field.

9 Looking to the future

Our subject area – science and the Indian tradition – has been considered from the perspectives both of history and of social science; sociology is, after all, as Ernst Troeltsch pointed out, the history of the present. In the introductory chapter the methodology was justified with reference to the work of Stanley J. Tambiah. The main arguments presented so far will now be summarized in order to indicate fertile areas for further research.

India, 'the next knowledge superpower', excels in many areas of science and technology – computers, medicine and nuclear power, for example – and these achievements are integral to national development at every level of society. This is not to say that such development might not be improved by adopting Amartya Sen's 'welfare economics'; the point is that India's science and technology are not merely for show. Amartya Sen's views were described in an earlier publication and it was noted briefly in the introductory chapter that India has an impressive record in relation to global environmental issues.[1]

It was important to acknowledge at the outset the various dimensions of the Indian tradition, of which the social aspect is often the most visible. This became clearer in the sociological material presented in Chapter 7, but has also been pointed out elsewhere (e.g. the temple vignette in Chapter 3). The book's sub-title, relating specifically to two great thinkers, might predispose the reader to consider the subject area purely in philosophical terms. That is not possible. The very fact that leading Brahmo Samajists expressed themselves differently depending on whether they were speaking in English or in Bengali should be a constant reminder of the relevance of social context!

The second chapter outlined the crucial encounter between European thought and educated India which took place in the nineteenth century. It was catalyzed by the decision in 1835 to use the English language as the medium of instruction in higher education. Whether we call the outcome a 'renaissance' or 'enlightenment' is immaterial; there was a reformulation of key elements of tradition which generated an irreversible momentum which paved the way for the eventual emergence of India, Pakistan and Bangladesh a century later. It is important to remember that a good deal of what has been described applies as much to Pakistan and Bangladesh as to India; it is only when dealing with the last half century that the scope of the study has been limited to India.

152 Looking to the future

The main features of the reform movements which occurred within the Hindu tradition and Islam have been characterized sometimes in terms of modernism, critical modernism and critical traditionalism (Bhikhu Parekh), and sometimes in terms of responses to secularization, noting the role of science within these processes. It was also recognized that educated Hindus, especially members of the Brahmo Samaj, focussed their beliefs increasingly in terms of *advaita* Vedānta, thus enabling them to counter the dualism of the West (especially of Christian missionaries) with a more integrated and spiritual monism. (This was facilitated towards the end of the nineteenth century by the publication of Thibaut's translation of Shankara's commentary on the Vedānta.)

A corollary of the renewed interest in *advaita* was that it enabled Indian reformers and scientists to interpret scientific advances from a Vedāntist perspective. As heat, light, sound, electricity and magnetism were progressively subsumed under common theories, culminating in Einstein's brilliant syntheses, *advaitins* saw the multiplicity of natural phenomena converging towards the One, that is *Brahman*. Scientists such as P. C. Roy and J. C. Bose interpreted their researches from this religious and philosophical perspective, though they conducted their various experiments and derived corresponding theories on the basis of Western methodology. They also saw connections between *pramāṇa*s such as *pratyakṣa*, translated as 'intuition', and the act of scientific discovery. These and other scientists also played an important role in the growth of national feeling, and were applauded for the quality of their work and their interpretation of it by leading figures such as Rabindranath Tagore, who was introduced at the end of the chapter.

If Chapter 2 considered the broad narrative of the nineteenth and early twentieth century, the third and fourth chapters addressed significant issues arising from it in more detail. Chapter 3 began by plunging the possibly reluctant reader into the Hindi-speaking world of a modern temple, with all its apparent paradoxes. The temple *āchārya* understood his Sanskrit texts very well, and excelled at chanting them, but equally important to him was his *lineage*, the line of succession back to the great Shankara himself. And what of his son, Subodh, on the one hand escorting his mother to the All India Medical Institute for the best laser eye treatment available anywhere in the world, while at the same time apparently believing in Rāhu, the eclipse demon? Subodh's interpretation of *sat-cit-ānanda*, based on his father's sermons, could form the basis for a new dialogue between Hindus and Christians. How many sons and daughters of Christian clergy in the West could expatiate as effectively on the significance of the Trinity?

Chapter 3 set Vedānta within the context of the six major schools, and offered a brief account of the main sources of scripture on which they are based. Scriptural shifts of emphasis, whereby, for example, the early Vedic sacrifices became interiorized, were discussed, and it was noted that environmental significance could be attributed to certain texts (e.g. Agni 'burning along... towards the east' interpreted in terms of settled cultivators clearing land as they moved from the Punjab into the Gangetic plains). A summary was given of the main features of the Vedāntic schools of Shankara and Rāmānuja, acknowledging certain common misunderstandings such as the tendency to translate *māyā* as something unreal.

It was observed that although Shankara's school became dominant in the late nineteenth century, and inspired scientists in their research, Rāmānuja's view that the whole world collectively and every being separately equals the 'body' of *Brahman* may give even more value to the material world.

Islam was considered, with particular reference to Syed Ahmad Khan and Muhammad Iqbal, in Chapter 2, but no further evaluation was attempted because this would have involved Pakistan and Bangladesh, which, from the mid-twentieth century onwards, are beyond our scope. References to Christianity in Chapter 2 were primarily in terms of the activities of missionaries. But by the turn of the century the churches were much more indigenous and it was possible to identify leading figures such as P. Chenchiah and Brahmabandhab Upadhyay. Paul Devanandan and M. M. Thomas came later.

Tagore's Hindu–Catholic friend Brahmabandhab Upadhyay set out arguments similar in some respects to those of Thomas Aquinas to establish the existence of a deity defined along lines compatible with *advaita* Vedānta. He illustrated the difference between religious truths of reason and of revelation in terms of the relationship between God's creative act and inner life, 'an inner life that revelation pronounces is trinitarian'. He answered the question of how far the light of reason can penetrate into divinity in terms of a conception of God as *sat-cit-ānanda*, the world being produced by *māyā*.

Upadhyay's preoccupation with Vedānta was not appreciated by the Roman Catholic Church of his day, though his theological work was consistent with his church's attempts to justify natural theology in terms of reconstructed Thomism. (Some of the nineteenth-century philosophical challenges to Christianity were discussed in the latter half of Chapter 4, where changes that were occurring in European science were also reviewed.) Upadhyay's work remains undervalued, and it is hoped that the occasion of the centenary of his death in 2007 will provide an opportunity to rectify this.

Chapter 4 appraised the development of science in India prior to the arrival of the British, and the state of science in Europe during the second part of the nineteenth century when the encounter between Western thought and educated India was most thoroughgoing. It was important to review the early history of science in India, partly because it falls under the rubric of our title, and partly to offset the contemptuous dismissal of Indian science by British administrators. It was also necessary to walk a tightrope in order to avoid the views of historians who have either adduced theories of Aryan dominance on the basis of flimsy or non-existent evidence, or have tried to attribute premature dates to Indian scientific achievements in order to avoid the possibility that some of them came from elsewhere.

European science was considered as far as possible from the perspective of educated India. Darwin's researches and evolution in general were important, but were never seen from the Indian perspective as inimical to religious belief. This, of course, raises questions as to why evolution caused such a furore in England; we concluded that a major issue for most people was the common ancestry presupposed between humans and animals. Just as new discoveries in geology and anthropology greatly expanded the nineteenth-century understanding of the extent of the universe in time, so advances in astronomy did as much for estimates of its extent in space. The progressive unification of the sciences under common heads was described, noting in particular certain notions, such as the ether, which Hindu Reformers and scientists thought to have special significance until it was dispensed with by Einstein and the experimental work of Michelson and Morley.

Einstein's major achievements were described in Chapter 5 together with the quantum views of Niels Bohr and the Copenhagen school. Reference was also made to collaborative work between Einstein and Satyendra Nath Bose which led to the genesis of Bose–Einstein statistics, at which point 'old' quantum theory was replaced by the new probabilistic approaches generally known as quantum mechanics. In 1927 Heisenberg's Uncertainty Principle proved that it is impossible fully to predict the behaviour of a microscopic system; this produced a rift between Einstein and his colleagues. But the consequent disagreements were not as clear cut as is often supposed; Heisenberg, for example, speculated that although quantum objects do not carry classical quantities such as position and momentum, they may carry the *potentiality* for such quantities. The chapter concluded with an analysis of Einstein's quest for a unified field (which fascinated a number of Indian scientists), and a brief account of where physics now stands. Throughout this chapter we resisted the views of those who found religious significance in new scientific discoveries, and endorsed Einstein's response to the Archbishop of Canterbury in 1921 that his theories were a purely scientific matter and had nothing to do with religion.

Chapter 6 considered the 'coming of age' of Indian science at both the institutional and the conceptual level. The work of P. C. Roy and J. C. Bose was evaluated in some detail, with emphasis on the manner in which they integrated their scientific research with their religious beliefs, and how Roy, for example, could express his scientific progress in terms of a combination of reason and intuition, which he attributed to the Brahmo Samaj. Bose focussed his research on the boundary lines between physics and physiology and investigated the possibility of pain in plants because that was in keeping with his understanding of *advaita* Vedānta. All the eminent Indian scientists that we considered in this chapter felt able to describe their scientific work to some extent in religious and philosophical terms, with the exception of Meghnad Saha, whose early mistreatment at the hands of caste-conscious brahmins had turned him against religion.

Chapter 7 provides an opportunity for young scientists to speak for themselves. Just under 700 completed questionnaires were obtained from four major centres, Delhi, Bangalore, Madurai and Kottayam, and 155 interviews were conducted. The data was used to test three hypotheses, two of which were found to require modification. The third, which was tested and established as valid according to standard statistical procedures, was as follows:

> The degree of perceived conflict between science and religion has an inverse correlation with the importance attached to religion. Thus a high degree of perceived conflict is related to a low valuation of religion, and a low degree of perceived conflict is related to a high valuation of religion.

The areas of potential conflict were chosen partly for historical reasons (e.g. biological evolution, theories of the universe's origin) and partly because preliminary discussions and four years of teaching physics at Delhi University suggested that they were becoming increasingly problematic (e.g. reincarnation). While some respondents stated that the study of science had diminished their beliefs, others had become more religious. Hindu and Christian respondents were divided into two groups and compared.

Chapter 8 evaluates Tagore and Einstein in more detail, and discusses the recorded conversations which took place between them in 1930. Tagore's major themes, as summarized by Amartya Sen, were considered. These were nationalism, traditionalism, educational commitment, freedom of the mind and interpretational epistemology; each area provided insights into his scientific and religious beliefs. It was acknowledged that Tagore's view of the supreme being is intimately bound up with his understanding of relationship as foundational to reality. The Supreme One is the ultimate guarantor of all relationships, and is love. Insofar that we are conscious of this truth of relationship, our human spirit can move beyond the realm of science. It is as a collective entity that humanity achieves its ideal.

Einstein's beliefs have been described as 'cosmic religion', and some of his statements endorsing this were cited. He admired Spinoza, and would have been encouraged to know that there are contemporary writers such as Stuart Hampshire who claim that Spinoza's philosophy is compatible with, and almost anticipates, current theories of evolution and genetics. The recorded conversations between Einstein and Tagore were discussed, and consideration was given to how they might have proceeded if a different scientific viewpoint had been expressed (e.g. by Niels Bohr). A group of scientists were invited to comment on the conversations from different faith perspectives, and it was concluded that there are good reasons for situating what has so far been a narrow Western discussion of the relationship between science and religion in an inter-faith context.

Avoiding the pitfalls

It is important to be clear about what is *not* being proposed in the previous chapters.

Meera Nanda castigates both the Hindu Right (e.g. the Vishwa Hindu Parishad) and its leftist counterpart for misrepresenting the relationship between science and tradition. She draws attention to a 'slick looking book' containing a 'warm, fuzzy and completely sanitised description of the faith', which offers Hindu ideas and topics at the secondary school level in the British school system. It advises British teachers to introduce Hindu *dharma* as 'just another name' for 'eternal laws of nature' first discovered by Vedic *guru*s, and subsequently confirmed by modern physics and biology.

All this is nonsense – though no more so than the creationist and Intelligent Design theories now so pervasive in the West – and, as Nanda correctly points out, is damaging to both science and genuine spirituality. But she also sees another school of contemporary thought, the 'postmodernist left', which is almost as objectionable:

> Postmodernist attacks on objective and universal knowledge have played straight into the Hindu nationalist slogan of all perspectives being equally true – within their own context and at their own level.[2]

Postmodernists are, of course, perfectly entitled to disagree with scientific triumphalism – and we saw plenty of evidence for that in the nineteenth century under the British Raj – but it does not follow that 'Western' science should be replaced by 'local traditions which are not entirely led by rational and instrumental criteria but make room for the sacred, the non-instrumental and even the irrational'.[3] Social constructionist theories of science complement very neatly the postmodernist mistrust of science by claiming that modern science is just one culture-bound way to look at nature; the content of all knowledge is socially constructed and the supposed facts of science reflect the dominant interests and cultural biases of Western societies and their supporting élites.

These views are also unacceptable, and it is ironic that the postmodernist leftists – to use Nanda's phrase – play so easily into the hands of their rightist opponents. But Nanda goes too far when she criticizes Vivekananda and the reform movements for presenting the Hindu tradition as the fulfilment of all science. The Reformers played a crucial role in reformulating and purifying a tradition with a lineage as distinguished as anything in Europe. It is as unrealistic to try to understand the beliefs of modern, educated Hindus without reference to them as it would be to try to comprehend contemporary Anglican Christianity without reference to the English Reformation. That some of Thomas Cranmer's theological views now seem in need of revision is to be expected; he none the less deserves a special place in history – and so do Swami Vivekananda and the Brahmo leaders.

Science is universal; there can be no 'Indian' science or 'Hindu' science. But there have been, and continue to be, a significant number of top-rank scientists in all the major fields of science belonging to a variety of religious communities – Hindus, Muslims, Christians, Sikhs and others – who maintain that their scientific work reflects their religious beliefs. Their contributions both as scientists and as religious believers need to be taken more seriously in order to achieve a deeper and more comprehensive understanding of the relationship between science and religion.

Hidden unity

Most scientific discoveries – even major ones such as the theories of relativity – have little or no significance for religious belief, properly understood. Some of them, especially in the biological sciences, may have important implications for ethics, however. In this section a few recent publications which reflect this general view will be summarized.

John C. Taylor's *Hidden Unity in Nature's Laws* offers a vivid account of physics in which he characterizes the search for a discovery of elegant laws that unify and simplify our understanding of the intricate universe in which we live.[4] Taylor – a mathematical physicist and former student of the Nobel Prize-winning Pakistani theoretician, Abdus Salem (1926–96) – says nothing in his book about religion or God, and yet his overall thesis endorses much of what we have been arguing in the previous chapters. Despite this, the closing chapter, possibly omitting the overoptimistic assessment of string theory, would repay careful scrutiny from an inter-faith perspective because it echoes the kind of philosophical worldview which motivated the Hindu scientists we encountered earlier.

John Bowker's *The Sacred Neuron* has more to say about religious belief, which he links to processes in the human brain. Steering a careful path between relativism and naïve realism, Bowker claims that facts exist independently of our opinions (Einstein would have approved of that). He calls these *conducive properties*, which are not coercive, but account for a high degree of unanimity among people in certain rational judgements and emotional responses. Thus although it is not correct to say that beauty is an inherent property of a circle (or the Apollo Belvedere), there are in the circle certain conducive properties that cause a common emotional response which includes a satisfying appreciation of beauty.

Bowker believes that brain cells (*neurons* or *glia* cells) respond to conducive properties in nature, resulting in culture-independent responses and human experiences. Where these involve religion, the integration of reason and emotion is fundamental. Where science is concerned, Bowker suggests that we must exercise critical realism:

> Critical realism means...that virtually everything we say, not least if we are scientists, is approximate, provisional, corrigible and often wrong, at least from the point of view of later generations. But (against deconstruction) it is at least wrong about something: there is sufficiently what there is in the case of the universe to act as a constraint on what we say, even though no one knows, incorrigibly and completely, 'what the universe is'. It is because there is consistent constraint and input from the external world (often in the form of conducive properties) that science achieves its reliability, while remaining at the same time corrigible.[5]

Critical realism is as important for religion as for science, and it is unfortunate that so many scientists and religious representatives persistently overstep their legitimate boundaries. Richard Dawkins does this, but it is heartening that he expends most of his vitriol on those who are just as bad in the advocacy of religion!

Bowker's understanding of the integration of reason and emotion resonates with the manner in which Brahmo Samajists and Hindu scientists interpreted scientific discovery in *advaitic* terms. This is a perfectly legitimate extension of a discussion which has hitherto been conducted without reference to non-Western traditions. It began in earnest with Michael Polanyi's 'tacit dimension' and

Arthur Koestler's biosociative thinking.[6] Might we not also describe the act of scientific discovery as the integration of reason and emotion along the lines of Shankara's *pramāṇa – pratyakṣa* (intuition), for example? And bearing in mind, as we have seen, that there are internal debates within both science and religion, might we not replace our *advaitic* terminology with words more conducive to Rāmānuja's school of thought?

The New Brain Sciences: Perils and Prospects, edited by Dai Rees and Steven Rose, brings together the thinking of some of the most distinguished scholars in the field.[7] Neuroscience, an all-inclusive term, embraces such disciplines as biology, biochemistry, molecular biology, neurology, psychology, psychiatry and philosophy, and relates to several more. But the book mentions religion only once. It is very much in the mould of John C. Taylor's *Hidden Unity*, in that it is full of metaphysical significance for believers in all religious traditions, without saying so.

Rees and Rose point out that research and development in the neurosciences has expanded so much in the last twenty years that some have dubbed the current decade 'the decade of the brain'. During this period our understanding of the brain and its functions has mushroomed. From genetics has come the identification of genes associated with 'normal' mental functions (e.g. learning and memory), and the genetic basis for dysfunctions that go with mental illnesses. From physics and engineering have come the new perceptions of the brain offered by imaging systems (e.g. functional magnetic resonance imaging), and from the information sciences come possibilities for modelling computational brain processes and imitating them on a computer. Some neuroscientists such as Crick believe that they are on the threshold of discovering the Holy Grail itself – the nature of consciousness!

But the mass of information now available is still too much to enable neuroscientists to formulate a coherent theory of consciousness. Most are committed to a psychophysical parallelism of brain and mind, but that does not solve problems in the area of brain states which philosophers refer to as 'qualia' or paradoxes relating to free will and genetic determinism.

From the point of view of religion, some of the promises of neurotechnology – better medicines, cures for loss of brain cells (e.g. Alzheimer's disease), and even stem cell technology – can be welcomed, but others have already proved deeply problematic. This is not the place to enter into a detailed discussion, though it is important to point out that in most European countries it is no longer acceptable to resolve ethical problems arising from scientific advances without reference to the views of representatives of more than one religious community. Part III of Rees and Rose's symposium deals with the legal aspects of neuroscience, but avoids the religious implications.

Even at the level of improved scientific understanding much of what is now known about the brain challenges conventional religious lore. Homosexuality, for example, is a condition for which neuroscientists still do not have a complete explanation, but brain imaging indicates that there may be a correlation with left-handedness, which everyone now accepts as natural. A fuller explanation may have to take into consideration new aspects of neuroscience and gene groupings, but it is only a matter of time before the total picture becomes clear.

Churches and religious bodies generally have totally disregarded the scientific understanding of homosexuality in their enthusiasm to promote what is essentially a political agenda against a minority. Not surprisingly, governments react by increasingly ignoring or bypassing religious groups when trying to estimate public opinion. On one occasion, when the British government set up a seminar to enable ethicists and religious representatives to suggest guidelines for the Human Genome Project, I was invited to participate. But not on behalf of my own church, which was represented by an array of cathedral canons, but on the recommendation of a Hindu professor at the university at which I was then teaching! Such participation, whether on behalf of one's own religious community or another one, will become less and less feasible unless the excesses of religious fundamentalists are curbed.

In the previous chapters we have referred at several points to the need for a better scientific understanding of consciousness, and in relation to Einstein's admiration of Spinoza we noted Stuart Hampshire's view that aspects of this way of thinking are compatible with, and almost anticipate, current theories of evolution and genetics. We also noted Amartya Sen's comparison between Tagore's understanding of the nature of reality and that of Hilary Putnam. This has no particular relevance to the biological sciences, though a striking sentiment by Donald M. Broom, professor of animal welfare in the University of Cambridge, does appear to give substance to Tagore's notion of a 'universal mind':

> We know from scientific information and from personal knowledge that any society is more than the sum of its parts. Hence it might be said that there is a spirit within the society. The concept of God as a spirit linking all sentient individuals is reconcilable with the biological background and usable by all.
>
> The meaning of God presented here is linked to existing and formerly existing sentient beings... God... started to have an impact after sentient beings had evolved and interacted significantly with one another. Hence it could be said that God was involved with the creation of the moral and social world.
>
> Each religion would be expected to have change-resisting qualities. These qualities are valuable when the net effect of the religion on individuals in the society affected by it is positive and the potential changes are damaging.... Since religions are fundamentally conservative and relatively unresponsive to change, they are vulnerable in times when there is a rapid influx of new knowledge.[8]

Donald Broom is writing explicitly about the evolution of morality and religion, but he does so against the background of non-human sentient beings (e.g. primates) whose existence has conventionally been downplayed by the Judaeo-Christian tradition at various stages of its history. Tagore and the representatives of the various religious traditions, whose views on the Einstein/Tagore conversations were adduced at the end of Chapter 8, would be very pleased with Donald Broom's endorsement of their positions!

Rees and Rose conclude their survey of the brain sciences by calling for partnerships between science and a range of public and private institutions (they do not mention churches), 'to work for the best futures, which can only be established by initiatives from all sides':

> The agenda would need to serve a double purpose: on the one hand to think out what the new insights of science might imply for our understanding of what it means to be human and, on the other, to explore the questions of ethics, law and social policy triggered by the new technologies.[9]

All religions have important things to say about what it means to be human, and considerable common ground exists between them. But they need to dialogue more effectively with the scientists whose new discoveries are forcing the pace of change. And within that dialogue representatives of the non-Western world must be invited to play a greater role.

Appendix A
The nature of reality

A conversation between Rabindranath Tagore and Albert Einstein in the afternoon of 14 July 1930, at the latter's residence in Kaputh:

Einstein (E.): Do you believe in the Divine as isolated from the world?

Tagore (T.): Not isolated. The infinite personality of Man comprehends the Universe. There cannot be anything that cannot be subsumed by the human personality, and this proves that the truth of the Universe is human truth. I have taken a scientific fact to illustrate this – Matter is composed of protons and electrons, with gaps between them; but matter may seem to be solid. Similarly humanity is composed of individuals, yet they have their interconnection of human relationship, which gives living solidarity to man's world. The entire universe is linked up with us in a similar manner, it is a human universe. I have pursued this thought through art, literature and the religious consciousness of man.

E.: There are two different conceptions about the nature of the universe: (1) The world as a unity dependent on humanity. (2) The world as a reality independent of the human factor.

T.: When our universe is in harmony with Man, the eternal, we know it as truth, we feel it as beauty.

E.: This is a purely human conception of the universe.

T.: There can be no other conception. This world is a human world – the scientific view of it is also that of the scientific man. There is some standard of reason and enjoyment which gives it truth, the standard of the Eternal Man whose experiences are through our experiences.

E.: This is a realization of the human entity.

T.: Yes, one eternal entity. We have to realize it through our emotions and activities. We realize the Supreme Man who has no individual limitations through our limitations. Science is concerned with that which is not confined to individuals; it is the impersonal human world of truths. Religion realizes these truths and links them up with our deeper needs; our individual consciousness of truth gains universal significance. Religion applies values to truth, and we know truth as good through our own harmony with it.

E.: Truth, then, or Beauty, is not independent of Man?

T.: No.

E.: If there would be no human beings any more, the Apollo of Belvedere would no longer be beautiful?

T.: No.

E.: I agree with regard to this conception of Beauty, but not with regard to Truth.

T.: Why not? Truth is realized through man.

E.: I cannot prove that my conception is right, but that is my religion.

T.: Beauty is in the ideal of perfect harmony which is in the Universal Being; truth the perfect comprehension of the Universal Mind. We individuals approach it through our own mistakes and blunders, through our accumulated experience, through our illumined consciousness – how, otherwise, can we know Truth?

E.: I cannot prove scientifically that truth must be conceived as a truth that is valid independent of humanity; but I believe it firmly. I believe, for instance that the Pythagorean theorem in geometry states something that is approximately true, independent of the existence of man. Anyway, if there is a *reality* independent of man there is also a truth relative to this reality; and in the same way the negation of the first engenders a negation of the existence of the latter.

T.: Truth, which is one with the Universal Being, must essentially be human, otherwise whatever we individuals realize as true can never be called truth – at least the truth which is described as scientific and can only be reached through the process of logic, in other words, by an organ of thoughts which is human. According to Indian Philosophy there is *Brahman* the absolute Truth, which cannot be conceived by the isolation of the individual mind or described by words, but can only be realized by completely merging the individual in its infinity. But such a truth cannot belong to Science. The nature of truth which we are discussing is an appearance – that is to say what appears to be true to the human mind and therefore is human, and may be called *māyā* or illusion.

E.: So according to your conception, which may be the Indian conception, it is not the illusion of the individual, but of humanity as a whole.

T.: In science we go through the discipline of eliminating the personal limitations of our individual minds and thus reach that comprehension of truth which is in the mind of the Universal Man.

E.: The problem begins whether Truth is independent of our consciousness.

T.: What we call truth lies in the rational harmony between the subjective and objective aspects of reality, both of which belong to the super-personal man.

E.: Even in our everyday life we feel compelled to ascribe a reality independent of man to the objects we use. We do this to connect the experiences of our senses in a reasonable way. For instance, if nobody is in this house, yet that table remains where it is.

T.: Yes, it remains outside the individual mind, but not outside the universal mind. The table which I perceive is perceptible by the same kind of consciousness which I possess.

E.: Our natural point of view in regard to the existence of truth apart from humanity cannot be explained or proved, but it is a belief which nobody can lack – no primitive beings even. We attribute to Truth a super-human objectivity; it is indispensable for us, the reality which is independent of our existence and our experience and our mind – though we cannot say what it means.

T.: Science has proved that the table as a solid object is an appearance, and therefore that which the human mind perceives as a table would not exist if that mind were naught. At the same time it must be admitted that the fact, that the ultimate physical reality of the table is nothing but a multitude of separate revolving centres of electric forces, also belongs to the human mind.

In the apprehension of truth there is an eternal conflict between the universal human mind and the same mind confined in the individual. The perpetual process of reconciliation is being carried on in our science and philosophy, and in our ethics. In any case, if there be any truth absolutely unrelated to humanity then for us it is absolutely non-existing.

It is not difficult to imagine a mind to which the sequence of things happens not in space, but only in time like the sequence of notes in music. For such a mind its conception of reality is akin to the musical reality in which Pythagorean geometry can have no meaning. There is the reality of paper, infinitely different from the reality of literature. For the kind of mind possessed by the moth, which eats that paper, literature is absolutely non-existent, yet for Man's mind literature has a greater value of truth than the paper itself. In a similar manner, if there be some truth which has no sensuous or rational relation to the human mind it will ever remain as nothing so long as we remain human beings.

E.: Then I am more religious than you are!

T.: My religion is in the reconciliation of the Super-personal Man, the Universal human spirit, in my own individual being. This has been the subject of my Hibbert Lectures, which I have called 'The Religion of Man'.

Rabindranath Tagore, *The Religion of Man*, Appendix II, London: George Allen & Unwin Ltd., 1931, pp. 222–5.

* * *

Appendix A

The following is an extract from a second conversation on 19 August 1930, also at Kaputh:

T.: I was discussing...today the new mathematical discoveries which tell us that in the realm of infinitesimal atoms chance has its play; the drama of existence is not absolutely predestined in character.

E.: The facts that make science tend towards this view do not say goodbye to causality.

T.: Maybe not; but it appears that the idea of causality is not in the elements, that some other force builds up with them an organised universe.

E.: One tries to understand how the order is on the higher plane. The order is there, where the big elements combine and guide existence; but in the minute elements this order is not perceptible.

T.: This duality is in the depths of existence – the contradiction of free impulse and directive will which works upon it and evolves an orderly scheme of things.

E.: Modern physics would not say they are contradictory. Clouds look one from a distance, but, if you see them near, they show themselves in disorderly drops of water.

T.: I find a parallel in human psychology. Our passions and desires are unruly, but our character subdues these elements into a harmonious whole. Are the elements rebellious, dynamic with individual impulse? And is there a principle in the physical world which dominates them and puts them into an orderly organisation?

E.: Even the elements are not without statistical order; elements of radium will always maintain their specific order, now and ever onwards, just as they have done all along. There is, then, a statistical order in the elements.

T.: Otherwise the drama of existence would be too desultory. It is the constant harmony of chance and determination which makes it eternally new and living.

E.: I believe that whatever we do or live for has its causality; it is good, however, that we cannot look through it.

Krishna Dutta and Andrew Robinson (eds) *Selected Letters of Rabindranath Tagore*, Cambridge: Cambridge University Press, 1997, p. 533.

Appendix B
Investigation questionnaire (Chapter 7)

The purpose of this investigation is to estimate the views of undergraduate and research science students in India concerning the relationship between science and religion. Where relevant, indicate your responses by placing a tick against answers you agree with.

Part A General background

1. Degree being undertaken
2. Subject
3. Year of course
4. Your age
5. Gender: male/female
6. Name and place of high school
7. Religious affiliation of high school (if any)
8. Name and place of previous college (if any)
9. Religious affiliation of previous college (if any)
10. Degree(s) obtained
11. Subject(s)
12. Intended career
13. Occupation of father

Part B Science and religion

In the following questions, the word *science* is used to mean not merely the results of scientific progress (e.g. nuclear power, improved fertilizers, and so on), but the principles and attitudes adopted by the scientist in any scientific research.

1. Whereas in the West science and religion have often been in conflict, this does not seem to have happened in India in spite of rapid scientific development. Do you agree/disagree?
   What explanation would you give?
2. Einstein has said, 'Science without religion is lame; religion without science is blind.' Do you agree with both parts of this statement?
   Both / first part only / second part only / neither
3. (a) Do you think that there is any conflict between your own religion and science?   Yes / No

(b) Has your degree course of study changed or modified your religious beliefs at all? Yes / No In what way?

4. Which, if any, of the following *do not* agree with a religious outlook? (Tick the parts where you feel there is conflict)
   (a) Biological evolution
   (b) Theories of universe's origin
   (c) Technological progress
   (d) The use of reason
   (e) The necessity for proof
   (f) Other..........

5. Which, if any, of the following *do not* agree with a scientific outlook?
   (a) Existence of the soul
   (b) Prayer
   (c) Life after death
   (d) Reincarnation
   (e) Miracles
   (f) Other..........

6. Indicate with a short statement how you think science and religion are related:

7. If you needed help on a *moral or ethical* issue related in some way to scientific progress (e.g. birth control, the use of nuclear weapons), to which *one* of the following sources would you be most likely to turn?
   (a) Philosopher
   (b) Scientist
   (c) Magazine articles
   (d) Religious leader
   (e) Politician
   (f) Other..........

8. Do you think that the future of the Hindu tradition in India will depend mainly on its ability to come to terms with scientific and industrial progress? Yes / No Why?

9. Which of the following predictions would you make for your religion?
   (a) It will disappear completely
   (b) It will merge with other religions
   (c) It will increasingly become a powerful social and political force
   (d) Social practices will disappear but religious beliefs will remain
   (e) There will be a revival of religious feeling
   (f) Other..........

10. Religion is important to me..........
    (a) At all times
    (b) Family occasions (e.g. weddings, funerals)
    (c) When I need help
    (d) Never

    Explain briefly the meaning of God to you.

11. Do you attend any kind of religious meetings or place of worship? Yes / No
    If yes, state the name of the religious society or place of worship.
    How frequently? Daily / weekly / fortnightly / monthly / special occasions only / annually

12. Do you think there is any contradiction between the following?
    (a) Astronomy and astrology Yes / No
    (b) Human rights, and the consequences of 'karma' Yes / No
    What do you understand by 'karma'?

Thank you for your cooperation.

Select glossary of terms

advaita Vedānta	the non-dual or monistic school of Vedānta (associated with Shankara)
ākāśa	space
antaryāmī	the inner controller
anubhava	experience, for Shankara the final realization beyond words and objects of experience
ārya	noble
ātman	self, spirit
bhadralok	the educated gentility in Bengal
bhakti	devotion
Brahman	literally, 'the Great One', the supreme spiritual being, ultimate reality
darśana	intellectual perspective or orientation, viewpoint
dharma	order, code of practice
guṇa	constituent of *prakṛti*, quality
jagat	the material world
jīva	the living self
jñāna	knowledge, consciousness
māyā	wondrous power, appearance
mokṣa	spiritual liberation
nirguṇa	without attributes
nirvāṇa	the extinction of attachments
niṣkāma karma	action without attachment, selfless work
pariṇāma	evolution
prakṛti	non-spiritual cosmogonic principle made up of three *guṇa*s
pramāṇa	method for acquiring valid knowledge
pratyakṣa	intuition
puruṣa	person or spirit
samādhi	the integrated state of mind
saṃsāra	cycle of existence, flow of life
sat-cit-ānanda	being, consciousness and bliss
smṛti	remembered tradition

śruti	canonical scripture
śūnya	empty, the void
sūtra	authoritative aphorism or text of aphorisms
tat tvam asi	'that *you* are', a key text in the Chāndogya Upanishad
Viśiṣṭādvaita	qualified non-dualism (associated with Rāmānuja)

Notes

1 Introduction

1 'India: the next knowledge superpower', *New Scientist*, Vol. 185, No. 2487, 19 February 2005, 30–53.
2 Indira Gandhi, 'Man and environment', 14 June 1972, in *Indira Gandhi on Environment*, Delhi: Government of India, Department of Environment, 1984, pp. 20–9.
3 David L. Gosling, 'Europe, not India, is the bigger polluter', *Independent*, 20 July 2005.
4 For a complete list, see David L. Gosling, *Religion and Ecology in India and Southeast Asia*, New York and London: Routledge, 2001 and New Delhi: Oxford University Press, 2001, pp. 181–7.
5 *Traditions, Concerns and Efforts in India, National Report to UNCED*, Delhi: Government of India, Ministry of Environment and Forests, June 1992.
6 Sunita Narain, editorial, *Down To Earth*, Delhi: Society for Environmental Communications, 30 November 2005, 6.
7 Parvathi Menon, 'For more women in science', *Frontline*, 30 December 2005, 39–42.
8 R. Ninian Smart, in John Bowker (ed.) *The Oxford Dictionary of World Religions*, Oxford: Oxford University Press, 1997, p. xxiv.
9 M. M. Thomas, *A Diaconal Approach to Indian Ecclesiology*, Rome: Centre for Indian and Inter-religious Studies, and Tiruvalla (Kerala): Christava Sahitya Samithy, 1995, p. 57.
10 Robert K. Merton, 'The institutional imperatives of science', in B. Barnes (ed.) *Sociology of Science*, London: Penguin, 1972, p. 66.
11 Isaac Pennington, quoted in Stephen Neill, *Christian Faith Today*, Harmondsworth, Middlesex: Penguin, 1955, p. 34.
12 Ṛgveda 5.53.9.
13 Janet M. Soskice, *Metaphor and Religious Language*, Oxford: Oxford University Press, 1985, p. 107.
14 John Hedley Brooke, *Science and Religion: Some Historical Perspectives*, Cambridge: Cambridge University Press, 1991.
15 Michael Robert Negus and Christopher Southgate, 'Some resources for theological thinking on God and the world from outside the Christian tradition', in Christopher Southgate (ed.) *God, Humanity and the Cosmos*, London and New York: T. and T. Clark International, 1999, revised and expanded 2005, pp. 242–59. Southgate cites Fritjof Capra, *The Web of Life*, London: Flamingo, 1996.
16 Keith Ward, *Religion and Creation*, Oxford: Oxford University Press, 1996.
17 *Frontline*, 2 June 2006, p. 11.
18 David L. Gosling, *Science and Religion in India*, Series on Religion, No. 21, Bangalore: Christian Institute for the Study of Religion and Society, and Chennai: Christian Literature Society, 1976.
19 Gosling, op. cit. (4), pp. 8–10.

20 Louis Dumont and David Pocock (eds) *Contributions to Indian Sociology*; see especially 1957, Vol. 1; 1959, Vol. 3; and 1960, Vol. 4.
21 Stanley J. Tambiah, *Buddhism and the Spirit Cults in North-east Thailand*, Cambridge: Cambridge University Press, 1970, p. 370.
22 Ibid., p. 374.

2 Science in India's intellectual renaissance

1 Deepak Kumar, *Science and the Raj: A Study of British India*, New Delhi: Oxford University Press, 1995, second edn 2006. Dhruv Raina, 'The unfolding of an engagement: "the dawn" of science, technical education and industrialization: India, 1896–1912', *Studies in History*, Vol. 9, No. 1, 1993, 87–117. Gyan Prakash, 'Science "gone native" in colonial India', *Representations*, Vol. 40, 1992, 154–78. Gyan Prakash has also contributed an important article on science in the Subaltern Studies series: 'Science between the lines', in Shahid Amin and Dipesh Chakrabarty (eds) *Subaltern Studies IX, Writings on South Asian History and Society*, New Delhi: Oxford University Press, 1996, pp. 59–83. For an assessment of archaeology in North India see Upinder Singh, *The Discovery of Ancient India: Early Archaeologists and the Beginnings of Archaeology*, New Delhi: Permanent Black, 2004.
2 Ram Mohan Roy, 'Letter to Lord Amherst', 1823, in W. T. de Bary *et al.* (eds) *Sources of Indian Tradition*, New York: Columbia University Press, 1958, reprinted in two volumes 1966, Vol. 2, p. 41.
3 Thomas Babington Macaulay, 'Minute on Education', in de Bary, ibid., p. 598.
4 Peter J. Bowler, *The Environmental Sciences*, Fontana History of Science, London: Fontana, 1992, pp. 310–11.
5 David L. Gosling, *Religion and Ecology in India and Southeast Asia*, New York and London: Routledge, 2001 and New Delhi: Oxford University Press, 2001, chapter 4.
6 Bhikhu Parekh, *Colonialism, Tradition and Reform*, New Delhi and London: Sage Publications, 1989, p. 35.
7 Ibid., p. 63.
8 Ibid., p. 45.
9 M. N. Srinivas, *Social Change in Modern India*, California: California University Press, 1968, p. 119.
10 Parekh, op. cit. (6), p. 60.
11 J. N. Farquhar, *Modern Religious Movements in India*, Delhi: Munshiram Manoharlal, 1947, p. 38.
12 *A Hundred Years of Calcutta University*, Calcutta, 1957, p. 36.
13 Y. Ramchandra, *Memoirs*; these are in Urdu and are available in the library of the Brotherhood of the Ascension in Delhi.
14 *Sambad Prabhakar*, editorials, 8 June, 22 July, 1847.
15 Alexander Duff, *India and Indian Missions*, Edinburgh, 1839, p. 616.
16 William Johnson, in *Proceedings of the Second Decennial Missionary Conference*, 1882, p. 124.
17 G. G. Gillan, in *Proceedings of the First Decennial Missionary Conference*, 1872, p. 121.
18 Keshub Sen, *Epistle to Indian Brethren*, in V. S. Naravane, *Modern Indian Thought*, London: Asia Publishing House, 1964, p. 15.
19 Farquhar, op. cit. (11), p. 118.
20 Dayanand Sarasvati, *Light of Truth*, New Delhi: Jan Gyan Prakashan, 1970, p. 79.
21 Vivekananda, *Complete Works*, Vol. V, Almora: Advaita Ashram, 1964, p. 519.
22 David L. Gosling, *Science and Religion in India*, Series on Religion, No. 21, Bangalore: Christian Institute for the Study of Religion and Society, and Chennai: Christian Literature Society, 1976, p. 50.
23 Ibid., p. 48.
24 Vivekananda, *Complete Works*, Vol. III, p. 341.

25 Ibid., Vol. II, p. 414.
26 Ibid., Vol. VI, p. 235.
27 I am grateful to Eric J. Sharpe for this version of Vivekananda's opening speech at the World's Parliament of Religions.
28 Vivekananda, *Raja Yoga: Conquering the Internal Nature*, London: Longmans, Green & Co., 1896.
29 Elizabeth de Michelis, *A History of Modern Yoga: Patañjali and Western Esotericism*, London: Continuum, 2004, p. 124.
30 Ibid., p. 150.
31 Sarah Strauss, *Positioning Yoga: Balancing Acts Across Cultures*, Oxford: Berg, 2005, p. 45.
32 Vivekananda, *Complete Works*, Vol. V, p. 298.
33 Ibid., p. 277.
34 Vivekananda, *Complete Works*, Vol. II, p. 440.
35 Ibid., p. 225.
36 G. H. Langley, *Sri Aurobindo*, Royal India and Pakistan Society, n.d.
37 Aurobindo, *The Life Divine*, Book 1, Pondicherry: Sri Aurobindo Ashram, 1970, p. 45.
38 Bradley-Birt, *Poems of Henry Derozio*, quoted in de Bary, op. cit. (2), p. 568.
39 Gosling, op. cit. (22), p. 38.
40 Syed Khan, in de Bary, op. cit. (2), p. 743.
41 Ibid., p. 744.
42 Muhammad Iqbal, 'Man and Nature', in *Poems from Iqbal*, London: John Murray, 1955, p. 3.
43 Muhammad Iqbal, *Reconstruction of Religious Thought in Islam*, in de Bary, op. cit. (2), p. 759.
44 Iqbal, op. cit. (42), p. 50.
45 Muhammad Iqbal, in Naravane, op. cit. (18), p. 293.
46 Muhammad Iqbal, in de Bary, op. cit. (2), p. 761.
47 L. S. S. O'Malley, *Modern India and the West*, Oxford: Oxford University Press, 1941, p. 242.
48 Francis Robinson, *Islam and Muslim History in South Asia*, New Delhi and Oxford: Oxford University Press, 2000; same author, *The 'Ulama of Farangi Mahall and Islamic Culture in South Asia*, London: C. Hirst, 2001. These books are reviewed in detail by Muhammad Qasim Zaman, *Journal of the Royal Asiatic Society*, Vol. 14, Part 3, November 2004, 253–63.
49 Rabindranath Tagore, trans. Nirad C. Chaudhuri, and quoted by him in 'Tagore: the true and the false', *Times Literary Supplement*, No. 3786, 27 September 1974.
50 Andrew Robinson, *Satyajit Ray: The Inner Eye*, London: Deutsch, 1989, p. 55.
51 Rabindranath Tagore, letter to Pratima Tagore, in Krishna Dutta and Andrew Robinson (eds) *Selected Letters of Rabindranath Tagore*, University of Cambridge Oriental Publications 53, Cambridge: Cambridge University Press, 1997, p. 83.
52 David Kopf, *The Brahmo-Samaj and the Shaping of the Modern Indian Mind*, Princeton: Princeton University Press, 1979, pp. 295–7.
53 Rabindranath Tagore, 'Brahmo Samajer Sarthakata', *Tattvabodhini Patrika* XVII, May 1911, 6–10.
54 E. J. Thompson, *Rabindranath Tagore, His Life and Work*, London: Oxford University Press, 1928, p. 44.
55 Rabindranath Tagore, 'The call to truth', *Modern Review*, Vol. XXII, No. 4, Calcutta, 1921, 429–33.
56 Julius J. Lipner, *Brahmabandhab Upadhyay: The Life and Thought of a Revolutionary*, New Delhi: Oxford University Press, 1999.
57 Rabindranath Tagore, letter to C. F. Andrews, 20 July 1926, quoted in Dutta and Robinson, op. cit. (51), p. 334.
58 Rabindranath Tagore, *The Religion of Man*, London: George Allen & Unwin, 1931, p. 206.

59 Rabindranath Tagore, *The Religion of Man*, p. 224.
60 Ibid., p. 20.
61 Robert B. Silvers and Barbara Epstein, *India: A Mosaic*, New York: The New York Review of Books, 2000, p. 67.
62 Kopf, op. cit. (52), p. 299.

3 Tradition redefined

1 Śatapatha Brāhmaṇa 1.4.1.14–17.
2 Julius J. Lipner, *Hindus: Their Religious Beliefs and Practices*, London and New York: Routledge, 1994, p. 156.
3 J. G. Suthren Hirst, *Śaṃkara's Advaita Vedānta: A Way of Teaching*, London and New York: Routledge Curzon, 2005, p. 14.
4 Lipner, op. cit. (2), p. 177.
5 John W. Bowker, *The Sacred Neuron: Extraordinary New Discoveries Linking Science and Religion*, London: I. B. Tauris & Co., Ltd., 2005, pp. 180, 50.
6 David L. Gosling, *A New Earth: Covenanting for Justice, Peace and the Integrity of Creation*, London: Council of Churches for Britain and Ireland, 1992, pp. 56–62.
7 Julius J. Lipner, *The Face of Truth: A Study of Meaning and Metaphysics in the Vedāntic Theology of Rāmānuja*, London: Macmillan, and Albany: State University of New York Press, 1986.
8 Hirst, op. cit. (3), pp. 49–52.
9 David L. Gosling, *Religion and Ecology in India and Southeast Asia*, New York and London: Routledge, 2001 and New Delhi: Oxford University Press, 2001, p. 110.
10 P. Chenchiah, editorial, *National Christian Council Review*, 1943, in Robin Boyd, *An Introduction to Indian Christian Theology*, Chennai: Christian Literature Society, 1969.
11 P. Chenchiah, editorial, *The Guardian*, Bangalore, 27 February 1947, in Boyd, ibid., p. 150.
12 Boyd, op. cit. (10), p. 188.
13 Ibid., p. 189.
14 P. Devanandan, 'Christian concern in Hinduism', in Boyd, op. cit. (10), p. 192.
15 P. Devanandan, 'Preparation for dialogue', in Boyd, op. cit. (10) , p. 196.
16 Julius J. Lipner, *Brahmabandhab Upadhyay: The Life and Thought of a Revolutionary*, New Delhi: Oxford University Press, 1999.
17 Rabindranath Tagore, *Cār Adhyāy*, Santiniketan: Sri Kisorimohan Satra (publisher), 1934, Preface, p. i.
18 Lipner, op. cit. (16), pp. 39, 41.
19 Ibid., p. 111.
20 Ibid., p. 114.
21 Ibid., p. 118.
22 Ibid., p. 116, footnote.
23 Job Kozhamthadam (ed.) *Contemporary Science and Religion in Dialogue: Challenges and Opportunities*, Pune: Association of Science, Society and Religion (ASSR) Publications, 2002; Job Kozhamthadam (ed.) *Science, Technology and Values: Science–Religion Dialogue in a Multi-religious World*, Pune: ASSR Publications, 2003; Job Kozhamthadam and Augustine Pamplany (eds) *East–West Interface of Reality*, Pune: ASSR Publications, and Aluva: Institute of Science and Religion, 2003; Job Kozhamthadam (ed.) *Religious Phenomena in a World of Science*, Pune: ASSR Publications, 2004; Sarojini Henry, *The Encounter of Faith and Science in Interreligious Dialogue*, Pune: Indian Institute of Science and Religion, and Delhi: ISPCK, 2005; Augustine Perumalil, *Critical Issues in the Philosophy of Science and Religion*, Pune: Indian Institute of Science and Religion, and Delhi: ISPCK, 2006.

Notes   173

24 G. Thibaut translated Shankara's commentary on the Vedānta in 1890: *The Vedāntic Sūtras of Bādarāyana with the Commentary by Śaṇkara*, Part I, Sacred Books of the East, trans. George Thibaut, ed. Max Müller, Vol. 34, Oxford: Clarendon Press, 1890.

4 Worldviews in encounter

1 S. Navaratna Rajaram and David Frawley, *Vedic Aryans and the Origins of Civilization: A Literary and Scientific Perspective*, New Delhi: Voice of India, 2001 (also published by World Heritage Press, Quebec).
2 Ibid., p. 58.
3 Bridget and Raymond Allchin, *The Birth of Indian Civilization*, Harmondsworth: Penguin, 1968, p. 281. For further information about Indian early history and prehistory, see M. K. Dhavalikar, *Man and Environment in Prehistory*, Symposia Papers, 4, Indian History Congress, Delhi: Amrit Printing, 1993. The bibliography contains references to additional material, some of which relates to early science.
4 Takao Hayashi, 'Indian mathematics', in Joseph W. Dauben (ed.) *The History of Mathematics from Antiquity to the Present: A Selective Annotated Bibliography*, New York: Garland, 1985, revised 2000.
5 A. N. Singh, 'Scientific thought in ancient India', in *History of Philosophy: Eastern and Western*, Vol. I, London: Allen & Unwin, 1952, p. 435.
6 Rajaram and Frawley, op. cit. (1), chapter 4, pp. 136–74.
7 A. Seidenberg, 'The origin of mathematics', in *Archive for the History of Exact Sciences*, Archive 18, New Delhi, 1978, p. 301.
8 Michio Yano, 'Calendar, astrology and astronomy', in Gavin Flood (ed.) *The Blackwell Companion to Hinduism*, Oxford: Blackwell, 2003, pp. 376–92.
9 Dominik Wujastyk, 'The science of medicine', in Flood, ibid., pp. 393–409.
10 Mircea Eliade, *Patterns in Comparative Religion*, New York: Sheed & Ward, 1958, p. 180.
11 Kauṭilya, *The Arthaśāstra*, in K. W. Morgan, *The Religion of the Hindus*, New York: Ronald Press, 1953, p. 101.
12 Megasthenes, *The Fragments*, 41, in H. Nakamura, *Ways of Thinking of Eastern Peoples*, Honolulu: East-West Center Press, 1966, p. 147.
13 A. L. Basham, *The Wonder that was India*, London: Sidgwick & Jackson, 1954, p. 499.
14 Severus Sebokht, in F. Nau, *Journal Asiatique*, Vol. II, 1910, pp. 225–7, quoted in Singh, op. cit. (5), p. 449.
15 Betty Heimann, *Facets of Indian Thought*, London: Allen & Unwin, 1964, p. 24.
16 Takao Hayashi, op. cit. (4).
17 Joseph Needham, *Heavenly Clockwork*, Cambridge: Cambridge University Press, 1960, p. 183.
18 W. E. Clark, 'Science', in G. T. Garrat (ed.) *The Legacy of India*, Oxford: Clarendon Press, 1937 (reprinted 1967), p. 345.
19 Heimann, op. cit. (15), pp. 37–9.
20 Singh, op. cit. (5), p. 446.
21 D. A. Somayaji, *A Critical Study of the Ancient Hindu Astronomy*, Dharwar: Karnatak University, 1971.
22 We are currently in the *kali yuga*, which began around 3000 BCE; when it ends, everything will be destroyed by flood and fire.
23 The significance of (3) may be seen by subtracting the second of the following two series from the first:

$$\infty = 1 + 2 + 3 + 4 + 5 + 6 + \ldots$$
$$\infty = 1 + 3 + 5 + 7 + 9 + 11 + \ldots$$

The result is an infinite series of even numbers. The Sanskrit term for infinity, *ananta*, literally means 'without end'.

24 Mañjula (b. 932 CE) made use of an infinitesimal increment to determine the true motion of planets, and Bhāskara knew that when a variable attains its maximum value, its differential vanishes. Sines and cosines were in use by the beginning of the fifth century, and Bhāskara eventually demonstrated that d/dx sin x = cos x = 0 for maximum values of x.
25 B. B. Dey, 'Scientific thought in ancient India: other sciences', in *History of Philosophy: Eastern and Western*, Vol. I, London: Allen & Unwin, 1952, p. 462.
26 Romila Thapar, *A History of India*, Vol. I, Harmondsworth: Penguin, 1966, p. 307. It is probable that this is a reference to Ash'arite theology. According to al-Ash'ari (b. 260 CE), space and time are both atomic and are defined as a series of infinitesimally small points separated from one another by an absolute void. Time is a 'succession of untouching moments and leaps across the void from one to the other with the jerk of the hand of the clock'.
27 Tahqiq-I-Hind Al-Bīrunī, in Romila Thapar, ibid., p. 239.
28 François Bernier, *Travels in the Mogul Empire A.D.1656–1668*, London: Constable, 1891; second edn, Delhi: S. Chand & Co., 1968, p. 339.
29 Ibid., pp. 339–40.
30 Ibid., pp. 243–5.
31 Zīa ad-dīn Barnī, in W. T. Bary et al. (eds) *Sources of Indian Tradition*, Vol. I, New York: Columbia University Press, 1958, p. 473.
32 Thapar, op. cit. (26), p. 303.
33 S. H. Nasr, *Science and Civilization in Islam*, Cambridge, MA: Harvard University Press, 1968, p. 228.
34 Basham, op. cit. (13), p. 499.
35 S. M. Ikram, *History of Muslim Civilization in India and Pakistan*, New York and London: Columbia University Press, 1964, p. 238. For a general overview see also Shireen Moosvi, *Man and Nature in Mughal Era*, Symposia Papers, 5, Indian History Congress, Delhi: Amrit Printing, 1993.
36 S. M. Ikram, ibid., p. 114.
37 Nasr, op. cit. (33), p. 59.
38 Martin Rudwick, *Bursting the Limits of Time: The Reconstruction of Geohistory in the Age of Revolution*, Chicago: Chicago University Press, 2005.
39 John Passmore, *A Hundred Years of Philosophy*, London: Duckworth, 1957, p. 35.
40 Alvar Ellegård, *Darwin and the General Reader: The Reception of Darwin's Theory of Evolution in the British Periodic Press 1859–1872*, Gothenburg: Gothenburg Studies in English, VIII, 1958, p. 43.
41 Benjamin Disraeli, *Lothair* (1870), in Passmore, op. cit. (39), p. 36.
42 A. R. Vidler, *The Church in an Age of Revolution*, Harmondsworth: Penguin, 1961, p. 117; O. Chadwick, *The Victorian Church*, Part II, London: Black, 1970, p. 9.
43 Vidler, op. cit. (42), p. 118.
44 Ibid., p. 117.
45 Herbert Spencer, *Principles of Sociology*, Vol. I, New York: Williams & Norgate, 1877, p. 106. Also quoted and discussed in *Critical Problems in the History of Science*, Madison: University of Wisconsin, 1959, p. 438.
46 Herbert Spencer, *First Principles*, Vol. II, New York, Williams & Norgate, 1911, p. 443.
47 John Baillie, *The Belief in Progress*, Oxford: Oxford University Press, 1950, p. 110.
48 S. C. Mukherji, 'The aim of science', *Modern Review*, April 1910.

5 Relativity and beyond

1 Andrew Robinson (ed.) *Einstein: A Hundred Years of Relativity*, Bath: Palazzo Editions Ltd, 2005, p. 232.
2 Ibid., p. 137.

Notes 175

3 Muhammad Iqbal, *Reconstruction of Religious Thought in Islam*, in W. T. de Bary et al. (eds) *Sources of Indian Tradition*, New York: Columbia University Press, 1958, p. 759.
4 Matthew Chalmers, 'Five papers that shook the world', *Physics World*, Vol. 18, No. 1, January 2005, 16–17.
5 Robinson, op. cit. (1), p. 89.
6 Albert Einstein, *The Principles of Relativity*, New York: Dover, 1923, p. 117.
7 Stephen Hawking, *The Universe in a Nutshell*, London: Bantam Press, 2001, p. 21.
8 George Gamow, *My World Line: An Informal Biography*, New York: Viking Press, 1970, p. 150.
9 Albert Einstein, letter to Max Born, 4 December 1926, in Max Born and Albert Einstein, *The Born–Einstein Letters*, London: Macmillan, second edn, 2005, quoted variously, including Robinson, op. cit. (1), p. 82.
10 John C. Polkinghorne, *The Quantum World*, New York and London: Longman, 1984, p. 53.
11 Ibid., p. 81.
12 Steven Weinberg, 'Einstein's search for unification', in Robinson, op. cit. (1), p. 107.
13 Freeman Dyson, preface, in Robinson, op. cit. (1), p. 9.
14 Roger Penrose, *The Road to Reality: A Complete Guide to the Laws of the Universe*, London: Jonathan Cape, 2004.
15 Ibid., p. 628.
16 Matthew Chalmers, 'Gravity's dark side', *Physics World*, London: Institute of Physics, June 2006, 18–21.
17 Ibid., p. 18.
18 John Polkinghorne, 'Beyond the big bang', in F. Watts (ed.) *Science Meets Faith*, London: SPCK, 1998, pp. 17–24; John D. Barrow and F. J. Tipler, *The Anthropic Cosmological Principle*, Oxford: Clarendon Press, 1986.
19 T. F. Torrance, in Max Jammer, *Einstein and Religion: Physics and Theology*, Princeton: Princeton University Press, 1999, p. 207.
20 Wolfhart Pannenberg, 'The doctrine of creation and modern science', *Zygon*, Vol. 23, 1988, 3–21.
21 The S-matrix is usually written as follows: $S_{if} = <f/i>$. Some of its analytic properties are still used (e.g. in relation to the veneziano amplitude).

6 Indian science comes of age

1 Jagadish Chandra Bose, 'Literature and science', in *Sir J. C. Bose, Life and Speeches*, Madras: Ganesh and Co., n.d., p. 80.
2 Charles E. Trevelyan, *On the Education of the People of India*, London: Longman and Orme, 1838, p. 31. Trevelyan is quoting from Princep's report.
3 Bhaskar Damodar, *The Student's Miscellany*, 1830, quoted in C. Heinsath, *Indian Nationalism and Hindu Social Reform*, Princeton: Princeton University Press, 1964, p. 13.
4 P. M. Holt, Ann K. S. Lambton and Bernard Lewis (eds) *The Cambridge History of Islam*, Vol. 2, Cambridge: Cambridge University Press, 1970, p. 115.
5 R. C. Majumdar, *British Paramountcy and Indian Renaissance*, Part II, Bombay: Bharatiya Vidya Bhavan, 1965, p. 53.
6 Prafulla Chandra Ray, *Life and Experiences of a Bengali Chemist*, London: Kegan Paul, Trench, Trübner and Co. Ltd., n.d., p. 200.
7 Ibid., p. 151.
8 P. C. Ray, *History of Hindu Chemistry*, Vol. 1, Bengal: Bengal Chemical and Pharmaceutical Works, 1902, p. 110.
9 Brajendranath Seal, *The Positive Sciences of the Ancient Hindus*, London: Longman and Green, 1915, p. 85. There may be an earlier issue of this book.
10 Ibid., p. 175.
11 G. K. Gokhale, *Speeches*, Calcutta: Natesan and Co., 1920, p. 309.

176 *Notes*

12 Roy, op. cit. (6), p. 42.
13 P. C. Roy, 'The discovery of mercurous nitrite', *Journal of the Asiatic Society*, Calcutta, 1896. Mercurous nitrite is extremely difficult to prepare because it is unstable. Its formula is Hg NO$_2$ or Hg – O – N = O. The NO$_2$ component 'wants to' become the very stable gas, nitric oxide, whereas mercury, which although a metal behaves more like a liquid, 'wants to' form into tiny globules. Even with sophisticated instruments it is extremely difficult to get it to separate as the nitrite.
14 Roy, op. cit. (6), p. 114.
15 Ibid., p. 166.
16 Ibid., p. 314.
17 Ibid., p. 99
18 Ibid., p. 541.
19 Ibid., pp. 30, 542.
20 Lord Hamilton, *Letter to Lord Elgin*, 12 November 1896, in Majumdar, op. cit. (5), p. 385.
21 Ashis Nandy, *Alternative Sciences: Creativity and Authenticity in Two Indian Scientists*, New Delhi: Oxford University Press, 1995, p. 23.
22 Ibid., p. 33.
23 Subrata Dasgupta, *Jagadis Chandra Bose and the Indian Response to Western Science*, New Delhi: Oxford University Press, 1999, p. 34.
24 J. C. Bose, *Modern Review*, February 1917, Vol. XXI, 202.
25 J. C. Bose, *Abyakto*, ed. Pulin Bihari Sen, Calcutta: Acharya Jagadis Chandra Bose Birth Centenary Celebration Committee, 1958, p. 87.
26 J. C. Bose, in M. Gupta, *Jagadish Chandra Bose, A Biography*, Bombay: Bharatiya Vidya Bhavan, 1964, p. 134.
27 Bose, op. cit. (1), p. 203.
28 Dasgupta, op. cit. (23), pp. 171, 255.
29 J. C. Bose, 'Voice of life', *Modern Review*, Vol. XXII, 590.
30 Nandy, op. cit. (21), p. 65.
31 Rabindranath Tagore, 'Acharya Jagadisher Jaivarta', *Vasudhara*, Vol. 2, 1957, 107.
32 P. C. Mahalanobis, quoted in S. Rangnathan, *Ramanujan, the Man and the Mathematician*, Bombay: Asia Books, 1967, p. 80.
33 Jawaharlal Nehru, *The Discovery of India*, Bombay: Asia Books, 1961, p. 221.
34 Interview conducted on 22 March 2006.
35 H. N. Verma and Amrit Verma, *Twenty-five Eminent Indians, 1947–2005*, Through the Ages Series No. 6, New Delhi: H. N. Verma, 2006, p. 120.
36 Ibid., p. 121.
37 William H. Cropper, *Great Physicists: The Life and Times of Leading Physicists from Galileo to Hawking*, New York: Oxford University Press Inc., 2001, p. 451.
38 Asutosh Mookerjee, *A Hundred Years of Calcutta University*, Calcutta: University of Calcutta, 1957, p. 239.

7 An investigation into the beliefs of Indian scientists

1 David L. Gosling, *Science and Religion in India*, Series on Religion, No. 21, Bangalore: Christian Institute for the Study of Religion and Society, and Chennai: Christian Literature Society, 1976.
2 For reasons of clarity these questions have been extracted from Appendix B.
3 Surajit Sinha (ed.) *Science, Technology and Culture*, New Delhi: Indian Research Council for Cultural Studies, 1970.
4 Ashok Parthasarathi, 'Sociology of science in developing countries: the Indian experience', *Economic and Political Weekly*, 2 August 1969, 1277. See also an article by A. D. King in the same publication, 29 August 1970, for an analysis of the social backgrounds of scientists at the Indian Institute of Technology in Delhi.
5 David L. Gosling, 'Planning for India's scientific development', *Physics Bulletin*, Vol. 21, 1970, 503–4; 'Scientists in Indian society', *South Asian Review*, Vol. 7, No. 4,

1974, 307–14; 'Scientific perspectives on rebirth', *Religion*, Vol. 4, No. 1, 1974; 'Some aspects of the impact of science on religion', *New Frontiers in Education*, Vol. 4, No. 3, 1974, 19–28; 'The Asian encounter between scientific thought and religious tradition', *Bangalore Theological Forum*, Vol. 7, No. 2, 1975, 21–34; 'Scientific decisions and social goals', in S. J. Samartha and L. de Silva (eds) *Man in Nature: Guest or Engineer?* Geneva: World Council of Churches, 1979, pp. 45–58; 'Religion and the environment', in D. J. R. Angell, J. D. Comer and M. L. N. Wilkinson (eds) *Sustaining Earth*, London: Macmillan, 1990, pp. 97–107; 'Science and Indian religion', *Religion and Society*, Vol. 45, No. 2, June 1998, 5–31.
6 David L. Gosling, 'Caste and sub-caste among Indian scientists', 'Scientific thought and religious tradition', 'Patterns of unbelief among Indian scientists' and 'Social patterns and the Indian scientific community'. Photocopies of these unpublished articles may be had on request.
7 'Examining reservation', *Frontline*, Vol. 23, No. 8, Chennai, 22 April 2006, 4–20.
8 Gosling, op. cit. (1), pp. 88–96.
9 Parthasarathi, op. cit. (4).
10 Daniel O'Connor, *Interesting Times in India: A Short Decade at St Stephen's College*, New Delhi: Penguin, 2005.
11 Gosling, op. cit. (1), and unpublished research paper.
12 For details of the Chi Square statistical test, see N. M. Downie and R. W. Heath, *Basic Statistical Methods*, New York: Harper & Row, 1965, pp. 164ff.
13 M. Enoch, *Evolution or Creation?* Chennai: Union of Evangelical Students of India, 1966.
14 The Adi Sanatan Deity Religion was founded in 1937 at Mount Abu in Rajasthan and has approximately 200 branches in India. It is an ascetic sect whose aims include self- and God-realization through yoga. The founder of the sect was a diamond merchant in Kolkata until, at the age of 60, he became the corporeal medium of Godfather Śiva. The ultimate aim is to re-establish a Golden Age of one government and a single religion. The sect believes that history is cyclical and repeats itself every 5000 years.
15 For further details of the arguments used by the Cārvākas against reincarnation see Ninian Smart, *Doctrine and Argument in Indian Philosophy*, London: George Allen and Unwin, 1964, p. 160.
16 See Sinha, op. cit. (3).

8 How clear is reason's stream?

1 Peter Reason, editor of *Action Research*, letter in *The Guardian*, 24 October 2005.
2 David L. Gosling, *Religion and Ecology in India and Southeast Asia*, New York and London: Routledge, 2001, and New Delhi: Oxford University Press, 2001, pp. 131–5.
3 Krishna Dutta and Andrew Robinson (eds) *Selected Letters of Rabindranath Tagore*, University of Cambridge Oriental Publications 53, Cambridge: Cambridge University Press, 1997. See also Amartya Sen, *Tagore and His India*, 21 November 2004, at http://www.countercurrents.org/culture-sen281003.htm. There are also some excellent publications on Tagore by William Radice.
4 Bhikhu Parekh, *Colonialism, Tradition and Reform*, New Delhi and London: Sage Publications, 1989, pp. 59–60.
5 David Kopf, *The Brahmo-Samaj and the Shaping of the Modern Indian Mind*, Princeton: Princeton University Press, 1979, p. 313.
6 Ibid., p. 295.
7 Dutta and Robinson, op. cit. (3), p. xxii.
8 Rabindranath Tagore, *The Religion of Man*, London: George Allen & Unwin Ltd., 1931, p. 191.
9 Ibid., p. 192.
10 Ibid., p. 206.
11 Ibid., pp. 134–5.

12. Rabindranath Tagore, *The Religion of Man*, p. 100.
13. Ibid., p. 102.
14. Ibid., p. 104.
15. Ibid., p. 165.
16. Ibid., p. 162.
17. Ibid., p. 15.
18. Ibid., pp. 141, 154.
19. Kopf, op. cit. (5), p. 309.
20. Dutta and Robinson, op. cit. (3), p. xxiv.
21. Rabindranath Tagore, *Gitanjali*, New Delhi: UBS Publications Ltd, 2003, p. 75.
22. Hilary Putnam, *The Many Faces of Realism*, Illinois: La Salle, 1987; see also Thomas Nagel, *The View from Nowhere*, New York, publisher not specified, 1989.
23. Dutta and Robinson, op. cit. (3), p. xxiii.
24. Nicholas Griffin (ed.) *The Selected Letters of Bertrand Russell, Volume I: The Private Years (1884–1914)*, London, publisher not specified, 1992, p. 417.
25. Albert Einstein, letters and papers, summarized in *The Guardian*, 22 May 2006.
26. Albert Einstein, 'Autobiographical notes', in Andrew Robinson (ed.) *Einstein: A Hundred Years of Relativity*, Bath, UK: Palazzo Editions Ltd., 2005, p. 31.
27. Ibid., p. 26.
28. Max Jammer, 'Einstein on religion, Judaism and Zionism', in Robinson, ibid., p. 185.
29. H. G. Kessler, *The Diary of a Cosmopolitan*, London: Weidenfeld & Nicolson, 1971, p. 322, quoted in Max Jammer, *Einstein and Religion: Physics and Theology*, Princeton: Princeton University Press, 1999, pp. 39–40.
30. Jammer, op. cit. (29), p. 48.
31. A. Reiser, *Albert Einstein – A Biographical Portrait*, New York: A. & C. Boni, 1930, p. 28, quoted in Jammer, ibid., p. 22.
32. Jammer, op. cit. (29), p. 11.
33. Ibid., p. 115.
34. Ibid., p. 73.
35. Ibid., p. 123.
36. Stuart Hampshire, *Spinoza and Spinozism*, Oxford: Clarendon Press, 2005.
37. Jammer, op. cit. (29), p. 129.
38. Ibid., p. 130.
39. Rabindranath Tagore, quoted in Mel Gussow, 'A meeting of two worlds', *Frontline*, 14 September 2001, 63.
40. Tagore, op. cit. (8).
41. Ibid., p. 206.
42. Quoted in Schlomith Flaum, 'At the feet of my master', *Palestine News*, 23 August 1941.
43. Dipankar Home and Andrew Robinson, 'Einstein and Tagore: man, nature and mysticism', *Journal of Consciousness Studies*, Vol. 2, No. 2, Summer 1995, 167–79; reproduced in a shortened and modified form in Dutta and Robinson, op. cit. (3), pp. 527–36.
44. Hilary Putnam, op. cit. (22).
45. Thomas Nagel, op. cit. (22).
46. Roger Penrose, *Shadows of the Mind: A Search for the Missing Sciences of Consciousness*, Oxford: Oxford University Press, 1994.
47. Werner Heisenberg, *The Physicist's Conception of Nature*, Hamburg, 1955; English translation, London: Hutchinson & Co., 1958, p. 15.
48. Lawrence Osborn, 'Theology and the new physics', in Christopher Southgate (ed.) *God, Humanity and the Cosmos*, London and New York: T. & T. Clark International, 2005, p. 135.
49. Penrose, op. cit. (46).
50. P. M. Holt, Ann K. S. Lambton and Bernard Lewis (eds) *The Cambridge History of Islam*, Vol. 2, Cambridge: Cambridge University Press, 1970, p. 778.
51. Ziauddin Sardar, 'Waiting for rain', *New Scientist*, 15 December 2001, 51.

9 Looking to the future

1 David L. Gosling, *Religion and Ecology in India and Southeast Asia*, New York and London: Routledge, 2001, and New Delhi: Oxford University Press, 2001, pp. 126–36.
2 Meera Nanda, 'Postmodernism, Hindu nationalism and "Vedic science"', *Frontline*, 2 January 2004, 78. See also same author, *Prophets Facing Backward: Postmodern Critiques of Science and Hindu Nationalism*, Piscataway, NJ: Rutgers University Press, 2004.
3 Nanda, 'Postmodernism, Hindu nationalism and "Vedic science"', 80.
4 John C. Taylor, *Hidden Unity in Nature's Laws*, Cambridge: Cambridge University Press, 2001.
5 John W. Bowker, *The Sacred Neuron: Extraordinary New Discoveries Linking Science and Religion*, London: I. B. Tauris Ltd, 2005, p. 126.
6 Michael Polanyi, *The Tacit Dimension*, London: Routledge and Kegan Paul, 1967.
7 Dai Rees and Steven Rose (eds) *The New Brain Sciences: Perils and Prospects*, Cambridge: Cambridge University Press, 2004.
8 Donald M. Broom, *The Evolution of Morality and Religion*, Cambridge: Cambridge University Press, 2003, p. 226.
9 Rees and Rose, op. cit. (7), p. 274.

Select bibliography

Barrow, John D. and Tipler, F. J., *The Anthropic Cosmological Principle*, Oxford: Clarendon Press, 1986.
Birch, L. Charles, *Regaining Compassion for Humanity and Nature*, Kensington, Australia: New South Wales University Press, 1993.
—— *On Purpose*, Kensington, Australia: New South Wales University Press, 1990.
Bowker, John W., *The Sacred Neuron: Extraordinary New Discoveries Linking Science and Religion*, London: I. B. Tauris & Co., Ltd., 2005.
Brooke, John Hedley, *Science and Religion: Some Historical Perspectives*, Cambridge: Cambridge University Press, 1991.
Dasgupta, Subrata, *Jagadis Chandra Bose and the Indian Response to Western Science*, New Delhi: Oxford University Press, 1999.
de Michelis, Elizabeth, *A History of Modern Yoga: Patañjali and Western Esotericism*, London: Continuum, 2004.
Deane-Drummond, Celia E., *Biology and Theology Today: Exploring the Boundaries*, London: SCM Press, 2001.
Dutta, Krishna and Robinson, Andrew (eds) *Selected Letters of Rabindranath Tagore*, University of Cambridge Oriental Publications 53, Cambridge: Cambridge University Press, 1997.
Gosling, David L., *Religion and Ecology in India and Southeast Asia*, New York and London: Routledge, 2001 and New Delhi: Oxford University Press, 2001.
—— 'Science and Indian religion', *Religion and Society*, Vol. 45, No. 2, June 1998, 5–31.
—— 'God in creation: towards an ecumenical and scientific consensus', *Bangalore Theological Forum*, Vol. XXVII, Nos. 1 and 2, March and June 1995, 17–27.
—— 'Redeeming creation', in M. Reuver, F. Solms and G. Huizer (eds) *The Ecumenical Movement Tomorrow*, Kampen: Kok Publishing House and Geneva: World Council of Churches, 1993, pp. 120–35.
—— *A New Earth: Covenanting for Justice, Peace and the Integrity of Creation*, London: Council of Churches for Britain and Ireland/Delta Press, 1992.
—— 'Perenungan Teologis dan Etis tentang Penerapan Teknologi bagi Pembangunan' (Theological and ethical reflections on the application of technology for development), in Supardan and Supardan (eds) *Ilmu, Teknologi dan Etika* (Science, Technology and Ethics), Jakarta: BKP, 1991, pp. 229–35.
—— 'Religion and the environment', in D. J. R. Angell, J. D. Comer and M. L. N. Wilkinson (eds) *Sustaining Earth*, London: Macmillan, 1990, pp. 97–107.
—— (ed.) *Creation and the Kingdom of God*, Church and Society Documents, Geneva: World Council of Churches, 1988.

—— (ed.) *Science and the Theology of Creation*, Church and Society Documents, Geneva: World Council of Churches, 1988.

—— (ed.) *Technology from the Underside*, Geneva: World Council of Churches and Quezon City: National Council of Churches in the Philippines, 1986.

—— 'Towards a credible ecumenical theology of nature', *Ecumenical Review*, Vol. 38, No. 3, July 1986, 322–32. First published as 'Auf dem Wege zu eine glaubwürdigen Theologie der Natur', *Ökumenische Rundschau*, Heft 2, April 1986, 129–44.

—— 'The nuclear debate in the United Kingdom and the contribution of the British Council of Churches', in F. Solms (ed.) *European Churches and the Energy Issue*, Heidelberg: Forschungsstatte der Evangelischen Studiengemeinschaft, 1980, pp. 118–25.

—— 'Scientific decisions and social goals', in S. J. Samartha and L. de Silva (eds) *Man in Nature: Guest or Engineer?* Geneva: World Council of Churches, 1979, pp. 45–58.

—— 'The morality of nuclear power', *Theology*, Vol. LXXXI, No. 679, January 1978, 25–32.

—— *Science and Religion in India*, Series on Religion, No. 21, Bangalore: Christian Institute for the Study of Religion and Society, and Chennai: Christian Literature Society, 1976.

—— 'The Asian encounter between scientific thought and religious tradition', *Bangalore Theological Forum*, Vol. 7, No. 2, 1975, 21–34.

—— 'Christian response within Hinduism', *Religious Studies*, Vol. 10, No. 4, 1974, 433–9.

—— 'Scientific perspectives on rebirth', *Religion*, Vol. 4, No. 1, 1974.

—— 'Scientists in Indian society', *South Asian Review*, Vol. 7, No. 4, 1974, 307–14.

—— 'Some aspects of the impact of science on religion', *New Frontiers in Education*, Vol. 4, No. 3, 1974, 19–28.

—— 'Planning for India's scientific development', *Physics Bulletin*, Vol. 21, 1970, 503–4.

Gosling, David L. and Davis, Howard (eds) *Will the Future Work?* Geneva: World Council of Churches, 1986.

Gosling, David L. and Montefiore, Hugh (eds) *Nuclear Crisis: A Question of Breeding*, Dorchester: Prism Press, 1977.

Gosling, David L. and Musschenga, Bert (eds) *Science Education and Ethical Values*, Geneva: World Council of Churches and Georgetown: Georgetown University Press, 1985.

Gosling, David L. and Sahadat, Lorraine M. J., 'Sahadat's universalist approach to mysticism', in Melchior Mbonimpa, Guy Bonneau and Kenneth-Roy Bonin (eds) *Mysticism: Select Essays: Essays in Honor of John Sahadat*, Sudbury, Ontario: Les éditions Glopro, 2002.

Henry, Sarojini, *The Encounter of Faith and Science in Inter-religious Dialogue*, Pune: Indian Institute of Science and Religion, and Delhi: ISPCK, 2005.

Hirst, J. G. Suthren, *Śaṃkara's Advaita Vedānta: A Way of Teaching*, London and New York: Routledge Curzon, 2005.

Home, Dipankar and Robinson, Andrew, 'Einstein and Tagore: man, nature and mysticism', *Journal of Consciousness Studies*, Vol. 2, No. 2, Summer 1995, 167–79.

Jammer, Max, *Einstein and Religion: Physics and Theology*, Princeton: Princeton University Press, 1999.

Kopf, David, *The Brahmo-Samaj and the Shaping of the Modern Indian Mind*, Princeton: Princeton University Press, 1979.

Kozhamthadam, Job (ed.) *Contemporary Science and Religion in Dialogue: Challenges and Opportunities*, Pune: Association of Science, Society and Religion (ASSR) Publications, 2002.

Kozhamthadam, Job and Pamplany, Augustine (eds) *East–West Interface of Reality*, Pune: ASSR Publications, and Aluva: Institute of Science and Religion, 2003.

Kumar, Deepak, *Science and the Raj: A Study of British India*, New Delhi: Oxford University Press, 1995, second edn 2006.

Lipner, Julius J., *Brahmabandhab Upadhyay: The Life and Thought of a Revolutionary*, New Delhi: Oxford University Press, 1999.

—— *Hindus: Their Religious Beliefs and Practices*, London and New York: Routledge, 1994.

—— *The Face of Truth: A Study of Meaning and Metaphysics in the Vedāntic Theology of Rāmānuja*, London: Macmillan, and Albany: State University of New York Press, 1986.

Nandy, Ashis, *Alternative Sciences: Creativity and Authenticity in Two Indian Scientists*, New Delhi: Oxford University Press, 1995.

Nasr, S. H., *Science and Civilization in Islam*, Cambridge, MA: Harvard University Press, 1968.

O'Connor, Daniel, *Interesting Times in India: A Short Decade at St Stephen's College*, New Delhi: Penguin, 2005.

Parekh, Bhikhu, *Colonialism, Tradition and Reform*, New Delhi and London: Sage Publications, 1989.

Penrose, Roger, *The Road to Reality: A Complete Guide to the Laws of the Universe*, London: Jonathan Cape, 2004.

—— *Shadows of the Mind: A Search for the Missing Sciences of Consciousness*, Oxford: Oxford University Press, 1994.

Polkinghorne, John, 'Beyond the big bang', in F. Watts (ed.) *Science Meets Faith*, London: SPCK, 1998, pp. 17–24.

—— *The Quantum World*, New York and London: Longman, 1984.

Prakash, Gyan, 'Science between the lines', in Shahid Amin and Dipesh Chakrabarty (eds) *Subaltern Studies* IX, Writings on South Asian History and Society, New Delhi: Oxford University Press, 1996, pp. 59–83.

—— 'Science "gone native" in colonial India', *Representations*, Vol. 40, 1992, 154–78.

Putnam, Hilary, *The Many Faces of Realism*, Illinois: La Salle, 1987.

Rangnathan, S., *Ramanujan, the Man and the Mathematician*, Bombay: Asia Books, 1967.

Robinson, Andrew (ed.) *Einstein: A Hundred Years of Relativity*, Bath: Palazzo Editions Ltd, 2005.

Robinson, Francis, *Islam and Muslim History in South Asia*, New Delhi and Oxford: Oxford University Press, 2000.

Rudwick, Martin, *Bursting the Limits of Time: The Reconstruction of Geohistory in the Age of Revolution*, Chicago: Chicago University Press, 2005.

Sahadat, John, *Ways to Meaning and a Sense of Universality*, Mississauga, Ontario: Canadian Educators' Press, 1998.

Sinha, Surajit (ed) *Science, Technology and Culture*, New Delhi: Indian Research Council for Cultural Studies, 1970.

Soskice, Janet M., *Metaphor and Religious Language*, Oxford: Oxford University Press, 1985.

Strauss, Sarah, *Positioning Yoga: Balancing Acts Across Cultures*, Oxford: Berg, 2005.

Tagore, Rabindranath, *Gitanjali*, New Delhi: UBS Publications Ltd, 2003.

—— *The Religion of Man*, London: George Allen and Unwin, 1931.

Ward, Keith, *Religion and Creation*, Oxford: Oxford University Press, 1996.

Yano, Michio, 'Calendar, astrology and astronomy', in Gavin Flood (ed.) *The Blackwell Companion to Hinduism*, Oxford: Blackwell, 2003, pp. 376–92.

Index

Abhishiktananda 44
Adi Sanatan Deity Religion 123
Aligarh University 25
Allchin, Bridget and Raymond 49
American College, Madurai 106
Amherst, Lord 13
Andrews, C. F. 29, 131
Anglicists 12
anthropic principle 82, 138
Aquinas, Thomas 45, 153
Aristotle 76, 102
Arthaśāstra 51–2
Arya 23
Arya Samaj 17, 24, 45, 122, 125
Āryabhaṭa 54
astrology 18, 51, 57
Aurobindo *see* Ghose, Aurobindo
Ayer, A. J. 120

Barrow, John D. 82
Barua, Ankur 147–8
Bergson, Henri 25–6
Bhabha, Homi J. 100
bhadralok 14, 30
Bharatiya Vidya Bhavan 21, 126
Bhavan's Journal 21, 112, 126
Birch, Charles L. 39
Bohm, David 142
Bohr, Niels 71, 75, 76, 141–2, 150, 154–5
Born, Max 74, 78, 141
Bose, Jagadish Chandra 9, 18, 22, 31, 66, 84, 87–8, 91–5, 101, 134, 137, 152, 154
Bose, Satyendra Nath 73, 77, 80, 89, 95, 101, 134, 154
Bowker, John 38–9, 157
Bowler, Peter J. 14
Boyd, Robin 42

Brahmo Samaj 15, 17, 88, 91, 122, 131, 149, 151, 154, 157
Brooke, John H. 8
Broom, Donald M. 159

Calcutta Medical College 16, 85
Capra, Fritjof 46
Caraka 50, 55
Carey, William 44
Cārvākas/Lokāyatas 35, 126
Chakrabarti, Bhismadev 145
Chandrasekhar, Subrahmanyan 99–100
Chatterjee, Bankim Chandra 28
Chenchiah, P. 42–4, 153
Cherian, Jacob 144
Chi Square 115–19, 123
Chipko movement 4
Christ's College, Cambridge 91–2
Church Missionary Society College, Kottayam 106, 107
creationism 4, 82, 144, 155
Crick, Francis 158
culture, definition of 6
Curie, Madame 95
Curzon, Lord 45, 86

Dalits 40
Darwin, Charles/Darwinism 15, 60–4, 110, 120, 122–4, 128–9, 133, 153
Davies, Paul 143
Dawkins, Richard 136, 143, 157
de Michelis, Elizabeth 20
Delhi College 16, 25, 31
Derozio, Henry 24, 30, 42, 85
Devanandan, Paul D. 42–4, 153
Du Chaillu 62
Duff, Alexander 16
Dumont, Louis 9, 106
Dutta, Krishna 130, 133

Index

Eddington, Arthur 99
Edwardes College, Peshawar 41, 85
Einstein, Albert 8, 11, 15, 26, 30, 60, 65–6, 68–83, 84, 86, 95, 101, 111, 129, 131, 133–4, 135–43, 152, 154–5, 157, 159
Eliade, Mircea 51
Elphinstone College, Bombay (Mumbai) 85
Eswaran, S. V. 10
ether 64, 84, 91, 101, 154
Evangelical Union 113, 121–3
evolution 60–4, 116, 146, 153

Faraday, Michael 65, 91
Frawley, David 48
Fry, Stephen 96

Galileo 76, 102
Gandhi, Indira 2, 3, 41, 133
Gandhi, Mohandas Karamchand 29, 126, 131, 134
genetically modified (GM) crops 2
Ghose, Aurobindo 14–15, 23–4
Gitāñjali (Tagore) 29, 133–4
Gokhale, G. K. 15, 88
Gora (Tagore) 28
Goyal, Supriya 145
Grant Medical College, Bombay 85
Griffiths, Bede 44
Gupta, Brijen 105

Hampshire, Stuart 137, 155, 159
Hardy, G. H. 96, 101
Hasan, Usama 143–4
Hawking, Stephen 72–3, 143
Heisenberg, Werner 75–6, 134, 142, 147, 154
Higgs boson 78, 80, 95
Hindu (Presidency) College, Calcutta (Kolkata) 24, 25, 84–5, 89, 95, 100
Hirst, J. G. Suthren 37
Home, Dipankar 141
homosexuality and science 158–9
Hoyle, Fred 79–80, 125
Huxley, T. H. 62, 66–7

Indian Institute of Science, Bangalore 86, 98, 100, 102–3, 106
Indian Institute of Technology, Delhi 128
Indus Valley Civilization 48–9
Intelligent Design 4, 82, 144, 155
Iqbal, Muhammad 25–7, 47, 70, 127, 153
Iqbal, Muzaffar 144

Jains 52–4, 108, 122
Jammer, Max 136
Jayasuriya, D. L. 105, 127
Jones, William 87

Kalam, A. P. J. 2, 100
Kant, Immanuel 19
Kauṭilya 51–2
Kew Gardens 14
Khan, Syed Ahmad 25, 31, 47, 153
Koestler, Arthur 158
Kozhamthadam, Job 46
Krishnamurthi, J. 120
Kumar, Deepak 12

Lahore Government College 25
Lipner, Julius J. 38, 44–6
Lokāyatas *see* Cārvākas

Macaulay, Thomas B. 13, 16, 30, 50
McTaggart, John 25–6, 45
Madras Christian College 41, 85, 120
Mahalanobis, P. C. 96, 134
Mandal Commission 106
Manu, laws of 18
Marshman, John C. 24
Maxwell, James C. 65, 68, 91
Mehta, Ajatshatru 146–7
Menuhin, Yehudi 136
Merton, Robert K. 6
Minkowski 66, 69, 72
Mīrābaī 34
Miranda House, Delhi 121
Mookerjee, Asutosh 100
Müller, Max 87

Nagel, Thomas 141–2
Nanda, Meera 155–6
Nandy, Ashis 91, 94
Narain, Sunita 4
Narayanan, K. R. 40
Narlikar, Jayant 79
Nasr, S. H. 58–9, 144
Needham, Joseph 53
Nehru, Jawaharlal 25, 40–1, 97
Newton, Isaac 59, 64, 67, 68, 71
Nietzsche, Friedrich 25
Nobel Prize 5, 29, 68, 70, 86, 88, 98–9, 157
nuclear power 2
Nyāya 35

Orientalists 13
Osborn, Lawrence 142

Index 185

Osmania University, Hyderabad 85
Owen, Wilfred 30

Pannenberg, Wolfhart 82–3
Parekh, Bhikhu 14, 152
Parthasarathi, Ashok 106, 107
Pennington, Isaac 6
Penrose, Roger 80, 141–2
Pippard, Brian 99
Planck, Max 70
Pocock, David 9, 106
Polanyi, Michael 157
Polkinghorne, John 39, 82, 143
Poona College of Engineering 85
Popper, Karl 60
Prakash, Gyan 12
Presidency College *see* Hindu (Presidency) College
Princep, James 85
Purāṇas 13
Pūrva-mīmāṃsā 35
Putnam, Hilary 134, 141–2, 159

Radhakrishnan, S. 43, 125–6
Rāhu 34, 50–1, 54, 57, 152
Raina, Dhruv 12
Rajaram, S. Navaratna 48
Ramakrishna 18, 34, 44–5, 110, 122, 126
Ramakrishna Mission 18, 31, 121–2, 124, 125
Raman, C. V. 86, 88–9, 98–9
Rāmānuja 30–1, 37–9, 46, 131, 137, 139, 152–3, 158
Ramanujan, Srinivasa 96–7, 101
Ramchandra, Yesudas 16, 31, 42
Ranganathananda 10, 113, 126
Ray, Prafulla Chandra 17, 40, 66, 86, 87–90, 100, 101, 152, 154
Ray, Satyajit 28, 133
Rees, Dai 158, 160
reincarnation 21–2, 104, 115–16, 123, 125, 128–9, 145–6, 155
Robinson, Andrew 130, 133, 141
Roorkee Engineering College 85
Rose, Steven 158, 160
Roy, Arundhati 42
Roy, M. N. 25
Roy, Ram Mohan 12, 15–17, 31, 104, 130
Rudwick, Martin 61
Russell, Bertrand 60, 112, 120, 134

Saha, Meghnad 25, 40, 97, 99–101, 134, 154
Sai Baba 10

St Stephen's College, Delhi 34, 41, 86, 91, 108, 144–6
St Stephen's Hospital, Delhi 41
St Xavier's College, Bombay 85
St Xavier's College, Calcutta 91
Śaiva-Siddhānta 35
Salem, Abdus 157
Samachar Darpan 16
Sambad Prabhakar 16
Sāṃkhya 19, 21–2, 35, 40, 43, 47, 52, 87, 93
Sangh Parivar 18, 31, 36
Sarasvati, Dayanand 17–18, 31, 36, 86, 127
Sardar, Ziauddin 144
Schleiermacher, Friedrich 137
Schrödinger, Erwin 70, 75, 147
Schrödinger's cat 76, 78, 142
Schweitzer, Albert 137
science, definition of 6, 136
Seal, Brajendranath 87, 100
Seidenberg, A. 49–50
Sen, Amartya 130–1, 133–4, 141, 149, 151, 155, 159
Sen, Keshub Chunder 17, 42, 44–5, 131
Shankara 11, 15, 19, 32–5, 37–40, 46, 137, 139, 143, 152–3, 158
Shantiniketan 28, 133
Sikhs 108, 124
Sinha, Purnima 105
Sinha, Surajit 105–6, 127
Smart, R. Ninian 5
S-matrix 81, 83
Soskice, Janet M. 7
Southgate, Christopher 8
Spencer, Herbert 63, 66–7
Spinoza, Baruch 135, 137, 150, 155, 159
Srinivas, M. N. 15
Standard Model 74, 78, 80, 81
Strauss, Sarah 21
string theory 71, 77, 157
Stuart, James 44
Sūfī 26–7, 56
śūnya 52–3
Suśruta 50, 55

Tagore, Debendranath 16–17, 131
Tagore, Rabindranath 8, 11, 15, 27–31, 70, 94, 101, 129, 130–5, 138–43, 145–50, 152, 155, 159
Tambiah, Stanley J. 9, 151
Tantra 34, 35
Tattvabodhini Patrika 16, 28
Taylor, John C. 157–8

Index

Taylor, Richard 42
Thapar, Romila 56, 58
Thomas, M. M. 6, 41–2, 153
Torrance, T. F. 82
Troeltsch, Ernst 151

Uberoi, Patricia 106
United Nations 3
United Theological College, Bangalore 42
Upadhyay, Brahmabandhab 24, 29, 32, 42, 44–7, 131, 153
Uttara-mīmāṃsā 35

Vaiśeṣika 35, 52, 55
Varahāmihira 53–5
Vedānta 35

Vedānta College, Calcutta 15
Vishwa Hindu Parishad 155
Vivekananda 14, 18–22, 31, 43, 44, 110, 122, 125–7, 156

Ward, Keith 8
Weinberg, Steven 77, 82
Williams, Monier 24
Wilson College, Bombay 85
Wittgenstein, Ludwig 60, 135, 148
World's Parliament of Religions 20

yoga 20–1
Yoga 35

Zīa ad-dīn Barnī 57

eBooks – at www.eBookstore.tandf.co.uk

A library at your fingertips!

eBooks are electronic versions of printed books. You can store them on your PC/laptop or browse them online.

They have advantages for anyone needing rapid access to a wide variety of published, copyright information.

eBooks can help your research by enabling you to bookmark chapters, annotate text and use instant searches to find specific words or phrases. Several eBook files would fit on even a small laptop or PDA.

NEW: Save money by eSubscribing: cheap, online access to any eBook for as long as you need it.

Annual subscription packages

We now offer special low-cost bulk subscriptions to packages of eBooks in certain subject areas. These are available to libraries or to individuals.

For more information please contact webmaster.ebooks@tandf.co.uk

We're continually developing the eBook concept, so keep up to date by visiting the website.

www.eBookstore.tandf.co.uk